Advanced Maths E
Core 2 for Edexcel

Welcome to Advanced Maths Essentials: Core 2 for Edexcel. This book will help you to improve your examination performance by focusing on all the essential maths skills you will need in your Edexcel Core 2 examination. It has been divided by chapter into the main topics that need to be studied. Each chapter has then been divided by sub-headings, and the description below each sub-heading gives the Edexcel specification for that topic.

The book contains scores of worked examples, each with clearly set-out steps to help solve the problem. You can then apply the steps to solve the Skills Check questions in the book and past exam questions at the end of each chapter. If you feel you need extra practice on any topic, you can try the Skills Check Extra exercises on the accompanying CD-ROM. At the back of this book there is a sample exam-style paper to help you test yourself before the big day.

Some of the questions in the book have a (◎) symbol next to them. These questions have a PowerPoint® solution (on the CD-ROM) that guides you through suggested steps in solving the problem and setting out your answer clearly.

Using the CD-ROM

To use the accompanying CD-ROM simply put the disc in your CD-ROM drive, and the menu should appear automatically. If it doesn't automatically run on your PC:

1. Select the My Computer icon on your desktop.
2. Select the CD-ROM drive icon.
3. Select Open.
4. Select core2_for_edexcel.exe.

If you don't have PowerPoint® on your computer you can download PowerPoint 2003 Viewer®. This will allow you to view and print the presentations. Download the viewer from http://www.microsoft.com

Pearson Education Limited
Edinburgh Gate
Harlow
Essex
CM20 2JE
England
www.longman.co.uk

First published 2005
ISBN   0 582 83666 2

Design by Ken Vail Graphic Design

Cover design by Raven Design

Typeset by Tech-Set, Gateshead

Printed in the U.K. by Scotprint, Haddington

The publisher's policy is to use paper manufactured from sustainable forests.

The Publisher wishes to draw attention to the Single-User Licence Agreement situated at the back of the book. Please read this agreement carefully before installing and using the CD-ROM.

We are grateful for permission from London Qualifications Limited trading as Edexcel to reproduce past exam questions. All such questions have a reference in the margin. London Qualifications Limited trading as Edexcel can accept no responsibility whatsoever for accuracy of any solutions or answers to these questions.

Every effort has been made to ensure that the structure and level of sample question papers matches the current specification requirements and that solutions are accurate. However, the publisher can accept no responsibility whatsoever for accuracy of any solutions or answers to these questions. Any such solutions or answers may not necessarily constitute all possible solutions.

1 Algebra and functions

1.1 The Factor Theorem

Use of the Factor Theorem.

For a polynomial $f(x)$, the **Factor Theorem** states that if $f(x) = 0$ when $x = a$, then $(x - a)$ is a factor of $f(x)$, and vice versa.

This can be written

$$f(a) = 0 \Leftrightarrow (x - a) \text{ is a factor of } f(x)$$

Note also that

$$f\left(\frac{a}{b}\right) = 0 \Leftrightarrow (bx - a) \text{ is a factor of } f(x)$$

Note:

The symbol $\Leftrightarrow$ means that the statement is true when read from left to right, or from right to left.

For example,

- if $f(3) = 0$, then $(x - 3)$ is a factor of $f(x)$
- if $f(-4) = 0$, then $(x + 4)$ is a factor of $f(x)$
- if $(x - 5)$ is a factor of $f(x)$, then $f(5) = 0$
- if $(2x + 3)$ is a factor of $f(x)$, then $f(-\frac{3}{2}) = 0$.

The Factor Theorem can be used to identify factors of $f(x)$ and hence solve $f(x) = 0$, as in the following example.

Example 1.1 It is given that $f(x) = x^3 + 4x^2 + x - 6$.

a Find the value of $f(1)$.

b Use the Factor Theorem to write down a factor of $f(x)$.

c Express $f(x)$ as a product of three linear factors.

d Hence solve the equation $x^3 + 4x^2 + x - 6 = 0$.

Step 1: Substitute the value into $f(x)$.

a $f(1) = 1^3 + 4 \times 1^2 + 1 - 6 = 1 + 4 + 1 - 6 = 0$

Step 2: Use the Factor Theorem.

b $f(1) = 0 \Rightarrow (x - 1)$ is a factor of $f(x)$.

Step 3: Write down the linear factor, find and factorise the quadratic factor.

c $f(x) = (x - 1)(ax^2 + bx + c)$

so $x^3 + 4x^2 + x - 6 \equiv (x - 1)(ax^2 + bx + c)$

Equate x^3 terms: $\quad x^3 = ax^3 \qquad \Rightarrow a = 1$

Equate constants: $\quad -6 = -c \qquad \Rightarrow c = 6$

Equate x^2 terms: $\quad 4x^2 = bx^2 - ax^2 \quad \Rightarrow 4 = b - a \Rightarrow b = 5$

Hence $f(x) = (x - 1)(x^2 + 5x + 6)$

$\qquad\qquad = (x - 1)(x + 2)(x + 3)$.

Note:

You could use algebraic division (see Section 1.3).

Step 4: Solve $f(x) = 0$ by the usual methods.

d $f(x) = 0 \Rightarrow (x - 1)(x + 2)(x + 3) = 0$

So $x = 1$, $x = -2$, $x = -3$.

If, in the above example, you had not been given a hint to find the first factor, then a good strategy would be to try numbers that are factors of the constant term. In this case, try factors of -6: $1, -1, 2, -2, 3, -3, 6$ and -6.

Factors of a cubic polynomial

A cubic polynomial, f(x), can be factorised into a linear and a quadratic factor. You may then be able to factorise further.

So f(x) will have one of the following:

- three distinct linear factors,

- repeated linear factors,

- a linear factor and a quadratic factor that cannot be factorised in real numbers.

Solving a cubic equation

When trying to solve f(x) = 0, factorise f(x) into a linear and a quadratic factor. Then try to factorise the quadratic factor.

If it does not appear to factorise, check the discriminant $b^2 - 4ac$, remembering that

- if $b^2 - 4ac \geqslant 0$, the quadratic equation can be solved by completing the square or using the quadratic formula,

- if $b^2 - 4ac < 0$, the quadratic equation will have no real solutions, and f(x) = 0 will have only one real solution.

Recall:
The discriminant
(C1 Section 1.4).

1.2 The Remainder Theorem

Use of the Remainder Theorem.

The **Remainder Theorem** states that, if a polynomial f(x) is divided by a linear term ($x - a$), the remainder will be f(a).

If f(x) is divided by $ax - b$ the remainder will be $f\left(\dfrac{b}{a}\right)$.

The Remainder Theorem is sometimes written

$$\frac{P(x)}{x - a} = Q(x) + \frac{P(a)}{x - a} \quad \text{or} \quad P(x) = (x - a)Q(x) + P(a)$$

where P(x) is a polynomial and Q(x) is another polynomial (known as the **quotient**) of one degree less than P(x).

Note:
f(x) will be restricted to quadratic and cubic polynomials.

Note:
The Factor Theorem is a special case of the Remainder Theorem, because if ($x - a$) is a factor of f(x), then f(a) = 0, indicating that $x = a$ is a root of f(x) = 0.

Example 1.2 Find the remainder when f(x) = $x^3 - 4x^2 - 5x + 7$ is divided by

a ($x + 2$)   **b** ($2x - 1$)

Step 1: Substitute the appropriate value of x into the polynomial.

a f(-2) = $(-2)^3 - 4(-2)^2 - 5(-2) + 7 = -8 - 16 + 10 + 7 = -7$
The remainder is -7.

b f($\frac{1}{2}$) = $(\frac{1}{2})^3 - 4(\frac{1}{2})^2 - 5(\frac{1}{2}) + 7 = \frac{29}{8}$

Tip:
Be careful with signs.

1.3 Algebraic division

Simple algebraic division.

Example 1.3 Divide $x^3 + 2x^2 - 3x + 7$ by $x - 2$.

The steps of this process are outlined on a PowerPoint slide Example 1.3 which can be found on the CD.

$$
\begin{array}{r}
x^2 + 4x + 5 \\
x - 2\overline{)x^3 + 2x^2 - 3x + 7} \\
\underline{x^3 - 2x^2} \\
4x^2 - 3x \\
\underline{4x^2 - 8x} \\
5x + 7 \\
\underline{5x - 10} \\
+17
\end{array}
$$

$$(x^3 + 2x^2 - 3x + 7) \div (x - 2) = x^2 + 4x + 5 + \frac{17}{x - 2}$$

Note: Algebraic division has the same structure as traditional long division with numbers.

Note: The remainder 17 can be confirmed by substituting $x = 2$ into the polynomial: $f(2) = 8 + 8 - 6 + 7 = 17$

Note: The final equation can also be written as $(x^3 + 2x^2 - 3x + 7)$ $= (x^2 + 4x + 5)(x - 2) + 17$

Example 1.4 $f(x) = x^3 + 3x^2 - 4x - 12$

Factorise $f(x)$ fully, given that $x = -3$ is a root of $f(x) = 0$.

$x = -3$ is a root of $f(x) = 0 \Rightarrow (x + 3)$ is a factor of $f(x)$.

Step 1: Divide the expression by the known linear factor.

$$
\begin{array}{r}
x^2 \qquad - 4 \\
x + 3\overline{)x^3 + 3x^2 - 4x - 12} \\
\underline{x^3 + 3x^2} \\
0 - 4x - 12 \\
\underline{- 4x - 12} \\
0
\end{array}
$$

Step 2: Factorise the quadratic factor.

$x^2 - 4 = (x + 2)(x - 2)$
$x^3 + 3x^2 - 4x - 12 = (x + 3)(x - 2)(x + 2)$

Note: Using algebraic division is an alternative to the method of identities described in Example 1.1.

Recall: $x^2 - 4$ factorises using the difference of two squares.

Example 1.5 Use the Factor Theorem to find a linear factor of $f(x)$ where $f(x) = 2x^3 - 5x^2 - x + 6$. Hence express $f(x)$ as a product of three linear factors.

Step 1: Find a value of x such that $f(x) = 0$.
Step 2: Use the Factor Theorem.

Try $x = 1$: $f(1) = 2 - 5 - 1 + 6 = 2 \neq 0$. Hence $(x - 1)$ is not a factor.
Try $x = -1$: $f(-1) = -2 - 5 + 1 + 6 = 0$.
Hence $(x + 1)$ is a factor of $f(x)$.

Step 3: Find and factorise the quadratic factor.

Divide $f(x)$ by $(x + 1)$:

$$
\begin{array}{r}
2x^2 - 7x + 6 \\
x + 1\overline{)2x^3 - 5x^2 - \quad x + 6} \\
\underline{2x^3 + 2x^2} \\
-7x^2 - x \\
\underline{-7x^2 - 7x} \\
6x + 6 \\
\underline{6x + 6} \\
0
\end{array}
$$

$2x^2 - 7x + 6 = (2x - 3)(x - 2)$

Step 4: State the three linear factors.

Hence $f(x) = (x + 1)(2x - 3)(x - 2)$.

Tip: Try factors of 6: 1, −1, ...

Example 1.6 When $f(x) = ax^3 + bx^2 + 4x - 6$ is divided by $(x - 1)$ the remainder is -1. When $f(x)$ is divided by $x - 3$ the remainder is 42. Find the values of a and b.

Step 1: Use the Remainder Theorem twice.

By the Remainder Theorem $f(1) = -1$.

Hence $a(1)^3 + b(1)^2 + 4(1) - 6 = -1$

$$a + b = 1 \qquad ①$$

Also by the Remainder Theorem $f(3) = 42$.

Hence $a(3)^3 + b(3)^2 + 4(3) - 6 = 42$

$$27a + 9b = 36 \qquad ②$$

Step 2: Solve the simultaneous equations formed.

$① \times 27 \qquad 27a + 27b = 27 \qquad ③$

$③ - ② \qquad \quad 18b = -9$

$$b = -\tfrac{1}{2}$$

Substitute into $①$ $\qquad a = \tfrac{3}{2}$

SKILLS CHECK 1A: Factor and Remainder Theorems

1 For the following polynomials, write down the roots of $f(x) = 0$.

    a $f(x) = (2x - 1)(x - 2)(3x + 1)$ **b** $f(x) = (x^2 - 4)(x^2 - 9)$

    c $f(x) = (4x - 1)(3x + 4)(x - 2)$ **d** $f(x) = (x + 3)(x - 1)^2$

2 Find the roots of the following, giving your answer in surd form, if appropriate.

    a $x^3 - 7x + 6 = 0$ **b** $x^3 + x^2 - 10x - 6 = 0$ **c** $x^3 + 6x^2 + 11x + 6 = 0$

3 Find the remainder when the given polynomial is divided by the given linear term.

    a $x^3 - 3x^2 + 2x - 1$ divided by $(x - 2)$.   **b** $x^3 + 7x^2 + 8x + 10$ divided by $(x + 1)$.

    c $3x^3 + x^2 - 1$ divided by $(3x + 1)$.   **d** $x^3 + 3x^2 - 4$ divided by $(x - 1)$.

    Explain the significance of the result in part **d**.

4 Divide the given polynomial by the given linear term.

    a $x^3 - 6x^2 + 4x - 3$ divided by $(x - 3)$.   **b** $x^3 + 6x^2 - x + 1$ divided by $(x + 1)$.

    c $2x^3 + x^2 - 4x + 5$ divided by $(x - 2)$.

5 Find the roots of the equation $x^3 - 6x^2 + x + 14 = 0$ given that 2 is a root. Give your answers in surd form if appropriate.

6 Use the Factor Theorem to find a linear factor of $P(x)$ where $P(x) = x^3 + 3x^2 + 3x + 1$. Hence express $P(x)$ as a product of three linear factors.

7 When the polynomial $f(x) = ax^3 + bx^2 + 3x + 4$ is divided by $(x + 1)$ the remainder is -4. When $f(x)$ is divided by $(x - 2)$ the remainder is 38. Find the values of a and b.

SKILLS CHECK 1A EXTRA is on the CD

1 $f(x) = 4x^3 + 3x^2 - 2x - 6$.

Find the remainder when $f(x)$ is divided by $(2x + 1)$.   [Edexcel Jan 2002]

2 **a** Using the factor theorem, show that $(x + 3)$ is a factor of $x^3 - 3x^2 - 10x + 24$.

   b Factorise $x^3 - 3x^2 - 10x + 24$ completely.   [Edexcel Nov 2002]

3 $f(x) = x^3 + ax^2 + bx - 10$, where a and b are constants.

When $f(x)$ is divided by $(x - 3)$, the remainder is 14.

When $f(x)$ is divided by $(x + 1)$, the remainder is -18.

   a Find the value of a and the value of b.   **b** Show that $(x - 2)$ is a factor of $f(x)$.
                                                                    [Edexcel June 2002]

4 $f(x) = px^3 + 6x^2 + 12x + q$.

Given that the remainder when $f(x)$ is divided by $(x - 1)$ is equal to
the remainder when $f(x)$ is divided by $(2x + 1)$,

   a find the value of p.

Given also that $q = 3$, and p has the value found in part **a**,

   b find the value of the remainder.   [Edexcel June 2003]

5 $f(x) = x^3 - 19x - 30$.

   a Show that $(x + 2)$ is a factor of $f(x)$.   **b** Factorise $f(x)$ completely.   [Edexcel Jan 2004]

6 A polynomial is given by $p(x) = 2x^3 + 3x^2 - 8x + 3$.

   a Find $p(1)$.   **b** Find $p(\frac{1}{2})$.

   c Hence, factorise $p(x)$ as a product of three linear factors.

7 A polynomial is given by $f(x) = 3x^3 + 2x^2 + qx + 6$, where q is a
constant. The remainder when $f(x)$ is divided by $(x - 5)$ is 14 times
the remainder when $f(x)$ is divided by $(x + 1)$.

   a Find the value of q.

   b Show that $(x + 3)$ is a factor of $f(x)$.

   c Factorise $f(x)$ as a product of linear factors.

8 It is given that $f(x) = x^3 + 8x^2 - 23x - 210$.

   a Show that $f(5) = 0$.

   b Use the factor theorem to write down a factor of $f(x)$.

   c Hence factorise $f(x)$ completely.

   d Solve the equation $f(x) = 0$.

9 The polynomial $p(x) = x^3 + ax^2 + bx + 5$ leaves a remainder of 1
when divided by $x + 1$ and leaves a remainder of 13 when divided by
$x - 2$. Find the values of the constants a and b.

Coordinate geometry in the (*x*, *y*) plane

2.1 The equation of a circle

The equation of a circle in the form $(x - a)^2 + (y - b)^2 = r^2$.

The general equation of a circle with centre (a, b) and radius r is $(x - a)^2 + (y - b)^2 = r^2$.

If the circle has centre $(0, 0)$ the equation is $x^2 + y^2 = r^2$.

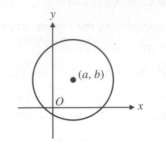

Note:
The brackets can be multiplied out to give the **expanded form** of the equation of a circle.

Example 2.1 **a** State the centre and radius of the circle
$(x + 3)^2 + (y - 4)^2 - 25 = 0$.

b Write the equation of the circle in expanded form.

Step 1: Rearrange the equation into the general form.

a $(x + 3)^2 + (y - 4)^2 - 25 = 0$
$(x - (-3))^2 + (y - 4)^2 = 5^2$
The centre is at $(-3, 4)$ and the radius is 5.

Step 2: Expand and simplify.

b $(x + 3)^2 + (y - 4)^2 - 25 = 0$
$x^2 + 6x + 9 + y^2 - 8y + 16 - 25 = 0$
$x^2 + y^2 + 6x - 8y = 0$

Tip:
This is a translation of $x^2 + y^2 = 25$ by -3 units in the *x*-direction and 4 units in the *y*-direction, which can be written $\begin{pmatrix} -3 \\ 4 \end{pmatrix}$.
See C1 Section 1.13.

Example 2.2 $A(3, -4)$ and $B(-5, 2)$ are the ends of a diameter of a circle. Find the equation of the circle.

Step 1: Draw a sketch.

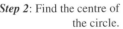

Tip:
Drawing a sketch will help you to see the situation and plan a strategy.

Step 2: Find the centre of the circle.

The centre of the circle, C, is at the middle of the diameter.

Midpoint of $AB = \left(\dfrac{3 + (-5)}{2}, \dfrac{-4 + 2}{2} \right) = (-1, -1)$

so the centre of the circle is at $(-1, -1)$.

Step 3: Find the radius.

Radius $= AC = \sqrt{(-1 - 3)^2 + (-1 - (-4))^2} = 5$

Step 4: Write the equation of the circle.

The equation of the circle is
$(x - (-1))^2 + (y - (-1))^2 = 5^2$
$(x + 1)^2 + (y + 1)^2 = 25$

Completing the square

Recall:
Completing the square
(see C1 Section 1.7).

You will recall from work on quadratic functions that

$$x^2 + 2ax = (x + a)^2 - a^2$$

You can use this to write the expanded form of the equation of a circle in the general form.

Example 2.3 A circle C has equation $x^2 + y^2 + 6x - 8y + 18 = 0$.

   a By completing the square, express this equation in the form
     $(x - a)^2 + (y - b)^2 = r^2$.

   b Write down the radius and the centre of the circle.

   c Describe a geometrical transformation by which C can be obtained
     from the circle with equation $x^2 + y^2 = r^2$.

Step 1: Rewrite, grouping the x-terms and y-terms.

Step 2: Complete the square for both x and y.

Step 3: Collect all constant terms on the right-hand side.

   a $x^2 + 6x + y^2 - 8y + 18 = 0$

     $(x + 3)^2 - 9 + (y - 4)^2 - 16 + 18 = 0$

          $(x + 3)^2 + (y - 4)^2 = 7$

   b The centre is at $(-3, 4)$ and the radius is $\sqrt{7}$.

Step 4: Compare with the general equation of a circle.

Step 5: Apply translation.

   c Translate O, the centre of the circle $x^2 + y^2 = 7$, three units to the
     left and four units up to obtain C, i.e. translate by the vector $\begin{pmatrix} -3 \\ 4 \end{pmatrix}$.

Recall:
Translations (C1 Section 1.13).

Example 2.4 Find the centre and radius of the circle $x^2 + y^2 = 10y$ and show the circle on a sketch.

Step 1: Rewrite, grouping the x-terms and y-terms.

$$x^2 + y^2 = 10y$$

$$x^2 + y^2 - 10y = 0$$

Step 2: Complete the square for both x and y.

Step 3: Collect all constant terms on the right-hand side.

$$x^2 + (y - 5)^2 - 25 = 0$$

$$x^2 + (y - 5)^2 = 25$$

Step 4: Sketch the circle. Hence the centre is at $(0, 5)$ and the radius is 5.

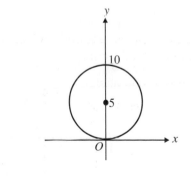

Tip:
Translate $x^2 + y^2 = 25$ by the vector $\begin{pmatrix} 0 \\ 5 \end{pmatrix}$, that is, 5 units up.

SKILLS CHECK **2A: The equation of a circle**

1 Give the general equation of the following circles.
   a Centre $(3, -2)$, radius 4   **b** Centre $(-5, 0)$, radius 5

2 Write the following circles in general form and hence state the centre and radius.
   a $x^2 + y^2 - 2x - 4y - 20 = 0$ **b** $x^2 + y^2 + 10x + 24y = 0$
   c $x^2 + y^2 - 6x + 10y + 18 = 0$ **d** $x^2 + y^2 + 2x - 6y + 3 = 0$

3 Find the equation of the circle with centre $C(2, -5)$ and point $A(6, -8)$ on the circumference.

4 Find the equation of the circle with diameter AB where A has coordinates $(4, -5)$ and B has coordinates $(-2, -3)$.

5 Find the centres and radii of the circles $x^2 + y^2 - 2x - 2y - 3 = 0$ and $x^2 + y^2 - 14x - 8y + 45 = 0$. Hence show that the circles touch each other.

6 The circle with equation $x^2 + y^2 - 4x - 8y + 7 = 0$ has centre C. The point $P(5, 2)$ lies on the circle.

   a Find the gradient of PC.

   b Find the equation of the line passing through P that is at right angles to the radius.

7 A circle C has equation $x^2 + y^2 = 4x - 6y + 3$.

   a Write the equation in the form $(x - a)^2 + (y - b)^2 = r^2$.

   b Write down the radius and the coordinates of the centre of the circle.

   c Describe a geometrical transformation by which C can be obtained from the circle with equation $x^2 + y^2 = r^2$.

8 A circle with equation $x^2 + y^2 = 49$ is translated 3 units down and 5 units to the right. Write down the equation of the translated curve in expanded form.

SKILLS CHECK **2A EXTRA** is on the CD

2.2 Coordinate geometry of a circle

Coordinate geometry of a circle using circle properties.

There are several circle properties that are useful in solving circle problems. Make sure that you can apply the following:

Circle property 1: angle in a semicircle

The angle subtended at the circumference of a circle by a diameter is a right angle.

Another way of saying this is: the angle in a semicircle is a right angle.

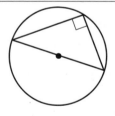

Example 2.5   **a** Show that the triangle ABC is right-angled, where A is $(0, -2)$, B is $(6, 6)$ and C is $(7, -1)$.

   b Given that A, B and C lie on the circumference of a circle, find the centre and radius of the circle and write down its equation.

Tip:
A fairly accurate sketch can be useful to help decide on a strategy to solve the problem. In this case it helps to identify where the right angle is.

Step 1: Draw a sketch.   **a** Using gradient $= \dfrac{y_2 - y_1}{x_2 - x_1}$

Step 2: Show that two sides of the triangle are perpendicular.

   Gradient $BC = \dfrac{-1 - 6}{7 - 6} = -7$

   Gradient $AC = \dfrac{-1 - (-2)}{7 - 0} = \dfrac{1}{7}$

   Since $\text{grad}_{BC} \times \text{grad}_{AC} = -1$, BC and AC are perpendicular and the triangle ABC is right-angled at C.

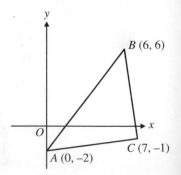

8

Step 3: Use circle properties to find the centre and radius.

b The angle in a semicircle is a right angle, so AB is a diameter of the circle. The centre of the circle is the midpoint of AB; call this M. The radius is given by the length of AM.

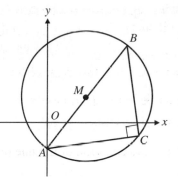

Coordinates of M

$$= \left(\frac{x_1 + x_2}{2}, \frac{y_1 + y_2}{2}\right)$$

$$= \left(\frac{0 + 6}{2}, \frac{-2 + 6}{2}\right)$$

$$= (3, 2)$$

$$AM = \sqrt{(3 - 0)^2 + (2 - (-2))^2} = 5$$

The circle has centre $(3, 2)$ and radius 5.
The equation of the circle is $(x - 3)^2 + (y - 2)^2 = 25$.

Recall:
The product of the gradients of perpendicular lines is -1 (C1 Section 2.2).

Note:
You could find AB^2, AC^2 and BC^2 and check Pythagoras' Theorem.

Example 2.6 $A(-5, -1)$, $B(7, 3)$ and $C(-1, 7)$ lie on a circle with diameter AB and centre P.

a Find the coordinates of P.

b Find the equation of the perpendicular bisector of AC, writing your answer in the form $ax + by + c = 0$, where a, b and c are integers.

c Show that the perpendicular bisector of AC passes through P.

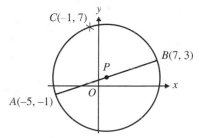

Step 1: Find the centre of the circle using circle properties.

a AB is a diameter, so P is at the midpoint of AB.

Coordinates of $P = \left(\frac{-5 + 7}{2}, \frac{-1 + 3}{2}\right) = (1, 1)$

Step 2: Put the additional information on the sketch.

Step 3: Find the equation of the perpendicular bisector of AC.

b Gradient of $AC = \frac{7 - (-1)}{-1 - (-5)} = \frac{8}{4} = 2$

Gradient of perpendicular bisector $= -\frac{1}{2}$

Midpoint of $AC = \left(\frac{-5 + (-1)}{2}, \frac{-1 + 7}{2}\right) = (-3, 3)$

The perpendicular bisector has gradient $-\frac{1}{2}$ and goes through $(-3, 3)$. The equation of the perpendicular bisector of AC is

$$y - 3 = -\frac{1}{2}(x - (-3))$$
$$2(y - 3) = -(x + 3)$$
$$2y - 6 = -x - 3$$
$$x + 2y - 3 = 0$$

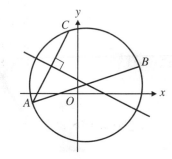

Recall:
The equation of the line through (x_1, y_1) with gradient m is $y - y_1 = m(x - x_1)$ (C1 Section 2.1).

Step 4: Check that the coordinates of P satisfy the equation of the line.

c When $x = 1$ and $y = 1$:

$$x + 2y - 3 = 1 + 2 - 3 = 0$$

Therefore $(1, 1)$ lies on the line $x + 2y - 3 = 0$ so the perpendicular bisector of AC passes through P.

Circle property 2: perpendicular to a chord

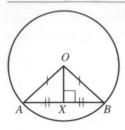

The perpendicular from the centre to a chord bisects the chord.

Note also:

- OAB is an isosceles triangle
- OX is the perpendicular bisector of chord AB.

Example 2.7 A circle with centre $C(1, 4)$ has chord AB where A is the point $(-4, 2)$ and B is the point $(3, k)$. Find the possible values of k.

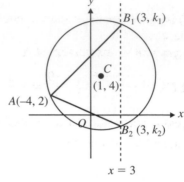

Step 1: Draw a sketch.

Note:
At this stage you don't know exactly where B is. But you do know it lies on $x = 3$, so there are two possible positions.

Step 2: Calculate the squares of the distances from C to A and B.

$$CA^2 = (1 - (-4))^2 + (4 - 2)^2 = 29$$
$$CB^2 = (1 - 3)^2 + (4 - k)^2 = 4 + 16 - 8k + k^2 = 20 - 8k + k^2$$

Recall:
Quadratic equations (C1 Section 1.7).

Step 3: Equate the distances and solve for k.

$CB = CA$ (both are radii) $\Rightarrow CB^2 = CA^2$

$$20 - 8k + k^2 = 29$$
$$k^2 - 8k - 9 = 0$$
$$(k - 9)(k + 1) = 0$$
$$k = 9 \text{ or } k = -1$$

Note:
The two values for k confirm the two possible positions for B of $(3, 9)$ and $(3, -1)$.

Example 2.8 AB is a chord of a circle. A is the point $(-2, 1)$ and B is the point $(4, -1)$. The centre of the circle is $C(2, k)$.

a Find the midpoint M of AB.

b Find the gradient of AB.

c Find the equation of the line MC.

d Hence, or otherwise, find the value of k.

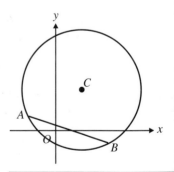

Step 1: Use the midpoint formula.

a $M = \left(\dfrac{-2 + 4}{2}, \dfrac{1 + (-1)}{2} \right) = (1, 0)$

Tip:
Show M on the sketch.

Step 2: Use the gradient formula.

b Gradient of $AB = \dfrac{-1 - 1}{4 - (-2)} = -\dfrac{1}{3}$

Step 3: Find the equation of the perpendicular bisector of AB.

c $AM = MB \Rightarrow CM$ is the perpendicular bisector of AB.

Gradient of $CM = 3$

Equation of CM:
$$y - 0 = 3(x - 1)$$
$$y = 3x - 3$$

Tip:
Use $m_1 \times m_2 = -1$ for perpendicular lines.

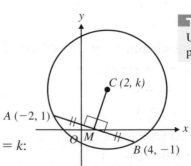

Step 4: Substitute point into the equation of the line.

d C lies on the line, so when $x = 2$ and $y = k$:
$$k = 3 \times 2 - 3 = 3$$
So $k = 3$.

10

Circle property 3: tangent to a circle

The tangent to a circle is perpendicular to the radius at its point of contact.

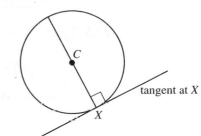

tangent at X

Example 2.9
$A(-1, 6)$ is a point on the circumference of a circle, centre $C(5, -3)$.
Find the gradient of the tangent at A.

Step 1: Draw a sketch.
Step 2: Find the gradient of the radius at A.

Gradient $CA = \dfrac{-3 - 6}{5 - (-1)} = \dfrac{-9}{6} = -\dfrac{3}{2}$

Step 3: Find the gradient of the line perpendicular to the radius.

The tangent is perpendicular to the radius
$\Rightarrow$ the gradient of the tangent is $\frac{2}{3}$.

$(-1, 6)\,A$

$\bullet C$
$(5, -3)$

Tip:
Leaving the gradient as a top-heavy fraction makes it easier to find the gradient of the perpendicular line.

Example 2.10
A circle has centre $C(-3, 2)$.
The line $y = x - 1$ is a tangent to the circle at A.

a Find the gradient of CA.

b Find the equation of CA.

c Find the coordinates of A.

C

$y = x - 1$

A

Step 1: Use circle properties.

a Gradient of tangent $= 1$
The radius is perpendicular to the tangent
$\Rightarrow$ gradient of $CA = -1$

Recall:
The gradient of the line $y = mx + c$ is m (C1 Section 2.1).

Step 2: Use the equation of a line $y - y_1 = m(x - x_1)$.

b Equation of CA:
$y - 2 = -1(x - (-3))$
$y - 2 = -(x + 3)$
$y - 2 = -x - 3$
$y = -x - 1$

Recall:
Product of gradients of perpendicular lines is -1 (C1 Section 2.2).

Step 3: Find the point of intersection of the lines by solving the simultaneous equations.

c A is the point of intersection of the lines
$y = x - 1$ ①
$y = -x - 1$ ②

Substituting for y from ① into ②
$x - 1 = -x - 1$
$2x = 0$
$x = 0$

Substituting for x in ①
$y = 0 - 1 = -1$

So the coordinates of A are $(0, -1)$.

Recall:
Solving two linear simultaneous equations (C1 Section 1.8).

1 $A(-4, 5)$, $B(4, 3)$ and $P(1, 0)$ are the vertices of triangle ABP.

   a Show that angle $APB = 90°$.

   b Hence show that P lies on the circumference of a circle with diameter AB.

   c The circle has centre C. Find the coordinates of C and the radius of the circle.

   d Hence write down the equation of the circle.

2 The point $(2, k)$ lies on the circumference of a circle with diameter AB, where A is the point $(-4, 1)$ and B is the point $(3, 2)$. Find the possible values of k.

3 $A(1, -2)$ and $B(-5, 4)$ are the ends of a chord of a circle and C is the point $(-1, 2)$.

   a Show that triangle ACB is isosceles.

   b Could C be the centre of the circle? Give a reason for your answer.

4 $A(-3, 3)$ and $B(5, 1)$ are the two ends of a chord AB of a circle.

   a Find the midpoint of AB.

   b Show, by using gradients, that $C(0, -2)$ could be the centre of the circle.

5 AB and CD are two chords of a circle, where A is $(1, 5)$, B is $(5, 3)$, C is $(3, -1)$ and D is $(5, 1)$.

   a Find the equation of the perpendicular bisector of AB.

   b Find the equation of the perpendicular bisector of CD.

   c Hence find the coordinates of the centre of the circle.

6 $M(1, -3)$ is the midpoint of a chord AB of the circle $x^2 + y^2 = 20$.
Find the equation of AB.

7 The point $P(5, 1)$ lies on a circle with centre $C(1, 4)$.

   a Show that P also lies on the line L with equation $3y - 4x + 17 = 0$.

   b Show that CP is perpendicular to L.

   c Deduce that L is a tangent to the circle at P.

SKILLS CHECK **2B EXTRA is on the CD**

The perpendicularity of radius and tangent.

A tangent and a normal can be drawn at any point on the circumference of a circle.

Since the radius is perpendicular to a tangent, the normal will pass through the centre of the circle.

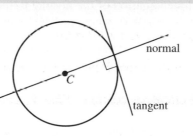

Note:
A normal is the line perpendicular to a tangent.

Example 2.11 Find the equation of the tangent to the circle, centre $C(4, 3)$, at the point $A(5, -2)$ on the circumference. Write your answer in the form $ax + by + c = 0$, where a, b and c are integers.

Step 1: Draw a sketch.

Gradient $CA = \dfrac{2 \quad 3}{5 - 4} = -5$

Step 2: Find the gradient of the radius and use it to find the gradient of the tangent.

Gradient of tangent $= \frac{1}{5}$

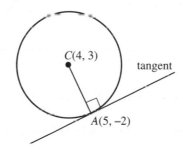

Step 3: Use $y - y_1 = m(x - x_1)$ to find the equation of the tangent.

Equation of tangent at A:

$y - (-2) = \frac{1}{5}(x - 5)$

$5(y + 2) = x - 5$

$5y + 10 = x - 5$

$x - 5y - 15 = 0$

Example 2.12 The point $(6, 8)$ is on the circumference of a circle with centre $C(2, 5)$. Find the equation of the normal to the circle at A.

Step 1: Draw a sketch.

The normal at A passes through C.

Step 2: Find the gradient of the radius.

Gradient $CA = \dfrac{8 - 5}{6 - 2} = \dfrac{3}{4}$

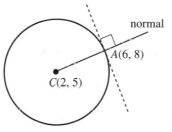

Tip:
You can use $C(2, 5)$ or $A(6, 8)$ as (x_1, y_1) when finding the equation of the line.

Step 3: Use $y - y_1 = m(x - x_1)$ to find the equation of the normal.

Equation of CA:

$y - 8 = \frac{3}{4}(x - 6)$

$4(y - 8) = 3(x - 6)$

$4y - 32 = 3x - 18$

$4y = 3x + 14$

SKILLS CHECK **2C: Tangents and normals to circles**

1 Find the equation of the tangent to the circle, centre $C(3, 5)$, at the point $A(1, 3)$.

2 Find the equation of the tangent to the circle $x^2 + y^2 + 4x - 6y + 8 = 0$ at the point $A(-1, 1)$.

3 Find the equation of the normal to the circle, centre $C(2, 2)$, at the point $A(5, 1)$.

4 Find the equations of the tangents to the circle, centre $C(-3, 2)$, at the points $A(1, 1)$ and $B(-4, -2)$. Hence find the point of intersection of the tangents.

5 $A(-8, -2)$ and $B(2, -2)$ are points on the circumference of a circle with centre $C(-3, \frac{1}{2})$. Find the equation of the normal at A and the equation of the tangent at B.

6 a Find the centre and radius of the circle $x^2 + y^2 - 4x + 10y + 4 = 0$ and hence sketch the circle.

b Show that $P(6, -2)$ lies on the circle.

c Find the equation of the normal to the circle at P, giving your answer in the form $ax + by + c = 0$, where a, b and c are integers.

7 $A(5, -1)$ and $B(7, 5)$ are two ends of a chord AB of a circle, centre $C(3, 3)$. Show that the tangents at A and B are perpendicular.

SKILLS CHECK **2C EXTRA** is on the CD

2.4 The intersection of a straight line and a circle

This section is concerned with the intersection of a straight line with a circle.

When finding the point(s) of intersection, you may have to solve a quadratic equation $ax^2 + bx + c = 0$. Here are the conditions relating to the discriminant of the quadratic polynomial $ax^2 + bx + c$. There are three cases to consider:

> **Recall:**
> The discriminant of a quadratic function (C1 Section 1.4).

1 If $b^2 - 4ac > 0$ there are two distinct real roots. This means there are two distinct points of intersection and the line cuts the circle twice.

2 If $b^2 - 4ac = 0$ there are two equal real roots, so there is only one point of intersection. The line touches the curve, so the line is a tangent to the circle at the point of intersection.

3 If $b^2 - 4ac < 0$ there are no real roots. The line and circle do not intersect.

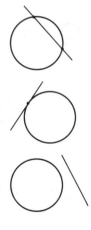

Example 2.13 Find the points of intersection of the circle $x^2 + y^2 = 20$ and the line $y = 3x - 10$.

Step 1: Substitute the linear equation into the equation of the circle and solve.

At the points of intersection, both these equations are satisfied.

$$y = 3x - 10 \qquad ①$$
$$x^2 + y^2 = 20 \qquad ②$$

> **Recall:**
> Solving simultaneous equations.

Substituting for y from ① into ②:

$$x^2 + (3x - 10)^2 = 20$$
$$x^2 + 9x^2 - 60x + 100 = 20$$
$$10x^2 - 60x + 80 = 0$$
$$(\div 10) \quad x^2 - 6x + 8 = 0$$
$$(x - 4)(x - 2) = 0$$
$$x = 4 \text{ or } x = 2$$

> **Note:**
> There may be two answers and they will be coordinates: there will be an x-value and a y-value.

> **Note:**
> For the quadratic equation $x^2 - 6x + 8 = 0$, $a = 1$, $b = -6$, $c = 8$.
> The discriminant:
> $b^2 - 4ac = 36 - 32 > 0 \Rightarrow$ there are two points of intersection.

Step 2: Substitute the values found in turn into the linear equation.

Substituting in ①:

when $x = 4$, $y = 3 \times 4 - 10 = 2$

when $x = 2$, $y = 3 \times 2 - 10 = -4$

Step 3: State the solution. The line and circle intersect at $(4, 2)$ and $(2, -4)$.

1 Find the point(s) of intersection of the following lines and circles.

 a $x^2 + y^2 = 25$ and $x + y = 7$

 b $x^2 + y^2 - 10x - 4y + 12 = 0$ and $y = x + 2$

 c $x^2 + y^2 - 4x - 2y - 15 = 0$ and $2x + y + 5 = 0$

2 Show that the line $y = x - 3$ is a tangent to the circle with centre $(-2, 1)$, radius $\sqrt{18}$, and find the point of contact.

3 Solve the following pairs of simultaneous equations, if possible, and interpret each result geometrically.

 a $(x - 3)^2 + (y + 4)^2 = 25$ and $y = 0$ **b** $x^2 + y^2 = 2x$ and $x + y = 3$

4 The straight line $y = x + 7$ intersects the circle $x^2 + y^2 = 25$ at the points P and Q.

 a Show that the x-coordinates of P and Q satisfy the equation $x^2 + 7x + 12 = 0$.

 b Hence find the coordinates of P and Q.

5 A circle, centre $C(4, 3)$, touches the line with equation $3x - 2y + 7 = 0$ at P.

 a Find the equation of CP.

 b By solving the equations simultaneously, find the coordinates of P.

 c Find the equation of the circle.

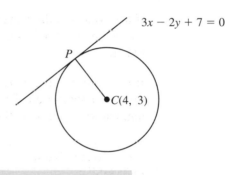

Examination practice   Coordinate geometry in the (x, y) plane

1 A circle C has equation

$$x^2 + y^2 - 10x + 6y - 15 = 0.$$

 a Find the coordinates of the centre of C.

 b Find the radius of C.   [Edexcel June 2001]

2 A circle C has equation

$$x^2 + y^2 - 6x + 8y - 75 = 0.$$

 a Write down the coordinates of the centre of C, and calculate the radius of C.

 A second circle has centre at the point $(15, 12)$ and radius 10.

 b Sketch both circles on a single diagram and find the coordinates of the point where they touch.
    [Edexcel June 2003]

3 A circle C has centre $(3, 4)$ and radius $3\sqrt{2}$. A straight line l has equation $y = x + 3$.

   a Write down an equation of the circle C.

   b Calculate the exact coordinates of the two points where the line l intersects C, giving your answers in surds.

   c Find the distance between these two points.   [Edexcel Jan 2002]

4 a Find in cartesian form an equation of the circle C with centre $(1, 4)$ and radius 3.

   b Determine, by calculation, whether the point $(2.9, 1.7)$ lies inside or outside C.
                                                                      [Edexcel Specimen Paper]

5 The points $A(-3, -2)$ and $B(8, 4)$ are at the ends of a diameter of the circle shown in the diagram.

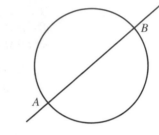

   a Find the coordinates of the centre of the circle.

   b Find an equation of the diameter AB, giving your answer in the form $ax + by + c = 0$, where a, b and c are integers.

   c Find an equation of the tangent to the circle at B.

   The line l passes through A and the origin.

   d Find the coordinates of the point at which l intersects the tangent to the circle at B, giving your answer as exact fractions.   [Edexcel June 2002]

6 A circle has cartesian equation $x^2 + y^2 - 4x + 6y - 12 = 0$. The centre of the circle is C and the radius is r.

   a Find the coordinates of C and the value of r.

   A straight line L has equation $4y + 3x = 19$. L intersects the circle at the point A.

   b Show that L is a tangent to the circle, and find the coordinates of the point A.

   The line L crosses the x-axis at the point B.

   c Find the area of triangle ABC.

7 The equation of a circle is $x^2 + y^2 - 2x + 6y - 15 = 0$.

   a Find the centre and radius of the circle.

   b The line $y = x + 1$ intersects the circle at A and B. Find the exact length of AB.

8 A circle C has equation $x^2 + (y - 5)^2 = 9$ and a straight line L has equation $y = \frac{4}{3}x$.

   a Write down:

   i the coordinates of the centre of the circle C,

   ii the radius of the circle C.

   b Show that the x-coordinate of a point common to C and L must satisfy the quadratic equation $25x^2 - 120x + 144 = 0$.

   c Solve this equation for x.

   d Comment on the geometrical relationship between C and L which is shown by your answer to **c**.

3 Sequences and series

3.1 Geometric series

The sum of a finite geometric series; the sum to infinity of a convergent geometric series, including the use of $|r| < 1$.

In C1, you looked at sequences and series, including arithmetic sequences and series.

The sequence 3, 6, 12, 24, 48, ... is defined by $u_n = 3(2^{n-1})$.

This is an example of a **geometric sequence**, where each term is formed from the previous one by *multiplying* by a constant, called the **common ratio** (r).

Adding the terms of a geometric sequence gives a **geometric series**, for example

$$3 + 6 + 12 + 24 + 48 + \cdots$$

Another example of a geometric series is $1 + 2 + 4 + 8 + 16 + \cdots$ because to get the next term we *multiply* the previous term by a constant, 2.
In this series, the first term is 1 and the common ratio is 2.

> **Recall:**
> The nth term can be denoted by u_n.

> **Tip:**
> To find r, divide any term by the previous term.

General expression for *n*th term, u_n

In a general geometric series, with first term a and common ratio r, the terms are as follows:

1st term $= a$
2nd term $= ar$
3rd term $= ar^2$
4th term $= ar^3$
.
.
.
nth term $= ar^{n-1}$

This an important formula for a geometric series, where you can quickly work out terms in the series.

> **Note:**
> The first term is usually denoted by a.

> **Tip:**
> You should learn the formula for the nth term.

Example 3.1   **a**   Find the nth term of the series given by:
$$2 + 6 + 18 + 54 + \ldots$$

b   Use your formula for the nth term to calculate the fifth term and the tenth term of this series.

Step 1: Identify the type of series and write the general series.

Step 2: Underneath write the information given in the question.

Step 3: Find any unknowns by solving the equations.

a   This is a geometric series because each term is multiplied by a constant to get the next term. The general series is:

$$a + ar + ar^2 + ar^3 + \cdots + ar^{n-1}$$
$$\downarrow \quad \downarrow \quad \downarrow \quad \downarrow$$
$$2 + 6 + 18 + 54 + \cdots$$

$a = 2$
$ar = 6 \rightarrow r = 3$

The nth term is $ar^{n-1} = 2(3^{n-1})$.

> **Tip:**
> Ensure the terms correspond.

b nth term $= 2(3^{n-1})$
Fifth term $= 2(3^4) = 162$; tenth term $= 2(3^9) = 39\,366$

Example 3.2 In January 2001, an investor put £1000 into a savings account with a fixed interest of 4% per annum. Interest is added to the account on 31 December each year and no further capital is invested.

a By what factor is the amount in the account increased when interest is added?

b How much will be in the account, to the nearest £, when interest has been added on 31 December 2015?

a This is a geometric series because each year you multiply by 1.04 to get the new amount. The general series is:

$$a \quad + \quad ar \quad + \quad ar^2 \quad + \cdots + ar^{n-1}$$
$$\downarrow \qquad\qquad \downarrow \qquad\quad \downarrow$$
$$1000(1.04) \; + \; 1000(1.04)^2 \; + \; 1000(1.04)^3 \; + \cdots$$

b $a = 1000(1.04)$, $r = 1.04$

You require the amount after 15 years, and to do this you must use the formula for the nth term:

nth term $= ar^{n-1}$, $n = 15$

Amount after 15 years $= 1000(1.04)^{14} = £1801$ (to the nearest £).

So, the amount in the account on 31 December 2015 is £1801.

Sum of the terms of a geometric series

The sum of the first n terms of a geometric series is:

$$S_n = \frac{a(1 - r^n)}{1 - r}$$

Proof

The sum of the first n terms is given by:

$$S_n = a + ar + ar^2 + \cdots + ar^{n-3} + ar^{n-2} + ar^{n-1} \qquad \text{①}$$

$$\times r$$

$$rS_n = \quad ar + ar^2 + \cdots + ar^{n-3} + ar^{n-2} + ar^{n-1} + ar^n \qquad \text{②}$$

Subtracting ① − ②: $\; S_n - rS_n = a - ar^n$

Factorise both sides: $\; S_n(1 - r) = a(1 - r^n)$

Dividing: $$S_n = \frac{a(1 - r^n)}{1 - r}$$

You can calculate the sum of the first four terms of the series $2 + 6 + 18 + 54 + \cdots$ using one of the versions of the formula for S_n, with $n = 4$, $a = 2$, $r = 3$:

$$S_4 = \frac{a(1 - r^n)}{1 - r} = \frac{2(1 - 3^4)}{1 - 3} = 80$$

Alternatively,

$$S_4 = \frac{a(r^n - 1)}{r - 1} = \frac{2(3^4 - 1)}{3 - 1} = 80$$

Example 3.3 The third term of a geometric series is $\frac{8}{3}$ and the sixth term is $\frac{64}{81}$. Find the first term, the common ratio and the sum of the first twenty terms.

Step 1: Write the general series.

You are told this is a geometric series. The general series is:

$$a + ar + ar^2 + ar^3 + \cdots + ar^5 + \cdots + ar^{n-1}$$

Step 2: Underneath write the information given in the question.

$$\downarrow \qquad\qquad\qquad \downarrow$$
$$\frac{8}{3} \qquad + \cdots + \qquad \frac{64}{81} \quad | \cdots$$

Step 3: Find any unknowns by solving the equation.

$ar^2 = \frac{8}{3}$ ①

$ar^5 = \frac{64}{81}$ ②

② divided by ① gives: $\dfrac{ar^5}{ar^2} = \dfrac{\frac{64}{81}}{\frac{8}{3}}$

$$\Rightarrow r^3 = \frac{8}{27}$$

$$r = \sqrt[3]{\frac{8}{27}} = \frac{2}{3}$$

Substitute into ①:

$$a\left(\frac{2}{3}\right)^2 = \frac{8}{3} \Rightarrow a = 6$$

Step 4: Use an appropriate formula.

You require the sum of the first twenty terms:

$$S_n = \frac{a(1 - r^n)}{(1 - r)}, \, n = 20, \, a = 6, \, r = \frac{2}{3}$$

Hence, $S_{20} = \dfrac{6\left(1 - \left(\frac{2}{3}\right)^{20}\right)}{1 - \frac{2}{3}} = 17.99$ (2 d.p.)

The sum to infinity

When the value of the common ratio lies between -1 and $+1$, the terms of the geometric series get smaller and smaller in size. In this case the sum of the series tends to a limiting value. This is known as the **sum to infinity** and is denoted by S_∞. The formula for this sum to infinity is:

$$S_\infty = \frac{a}{1 - r} \text{ provided } -1 < r < 1 \text{ (or } |r| < 1)$$

Example 3.4 For the following geometric series, find the nth term and also the sum to infinity:

$$1 - \frac{1}{2} + \frac{1}{4} - \frac{1}{8} + \cdots$$

Step 1: Write the general series.

You are told this is a geometric series. The general series is:

$$a + ar + ar^2 + ar^3 + \cdots + ar^5 + \cdots + ar^{n-1}$$

Step 2: Underneath write the information given in the question.

$$\downarrow \quad \downarrow \quad \downarrow \quad \downarrow$$
$$1 + \left(-\frac{1}{2}\right) + \frac{1}{4} + \left(-\frac{1}{8}\right) + \cdots$$

Step 3: Find any unknowns by solving the equations.

Clearly, $a = 1$, $r = -\frac{1}{2}$

Step 4: Use an appropriate formula.

nth term $= ar^{n-1} = \left(-\frac{1}{2}\right)^{n-1}$

We can calculate the sum to infinity because $-1 < r < 1$.

$$S_\infty = \frac{a}{1 - r} = \frac{1}{1 - \left(-\frac{1}{2}\right)} = \frac{2}{3}$$

$\sum$ notation

A geometric series can be written using $\sum$ notation.

Tip:
When r is a power then the series will be geometric.

Example 3.5 Evaluate $\displaystyle\sum_{r=1}^{9} 3(2^r)$.

Step 1: Expand the series and identify a, r and n.

$$\sum_{r=1}^{9} 3(2^r) = 3(2^1) + 3(2^2) + \cdots + 3(2^9)$$

This is a geometric series with $a = 3 \times 2 = 6$, $r = 2$, $n = 9$. S_9 is required.

Note:
Be careful: the r used in the summation is different from the common ratio, r.

Step 2: Use appropriate geometric series formula for sum.

Using $S_n = \dfrac{a(1 - r^n)}{1 - r}$,

$$\sum_{r=1}^{9} 3(2^r) = S_9 = \frac{6(1 - 2^9)}{1 - 2} = 3066$$

Example 3.6 Find $\displaystyle\sum_{r=3}^{9} 2^r$.

Step 1: Expand the series and identify a, r and n.

$$\sum_{r=3}^{9} 2^r = 2^3 + 2^4 + 2^5 + \cdots + 2^9$$
$$a = 2^3, r = 2, n = 7$$

Tip:
It is useful not to simplify the terms, so that you can see the common ratio more easily.

Step 2: Use appropriate geometric series formula for sum.

Using $S_n = \dfrac{a(1 - r^n)}{1 - r}$, with $n = 7$, $a = 8$, $r = 2$,

$$\sum_{r=3}^{9} 2^r = S_7 = \frac{8(1 - 2^7)}{1 - 2} = 1016$$

Note:
n = top number of the summation minus the bottom number plus 1.

Example 3.7 Evaluate $\displaystyle\sum_{r=0}^{\infty} \left(\tfrac{1}{2}\right)^r$.

Step 1: Expand the series and identify a, r and n.

$$\sum_{r=0}^{\infty} \left(\tfrac{1}{2}\right)^r = \left(\tfrac{1}{2}\right)^0 + \left(\tfrac{1}{2}\right)^1 + \left(\tfrac{1}{2}\right)^2 + \cdots = 1 + \tfrac{1}{2} + \tfrac{1}{4} + \cdots$$

This is a geometric series with $a = 1$, $r = \tfrac{1}{2}$. S_∞ is required.

Tip:
The sum to infinity is required.

Step 2: Use appropriate geometric series formula for sum.

$$\sum_{r=0}^{\infty} \left(\tfrac{1}{2}\right)^r = S_\infty = \frac{a}{1 - r}$$
$$= \frac{1}{1 - \tfrac{1}{2}}$$
$$= 2$$

Tip:
Check that $|r| < 1$ is satisfied.

SKILLS CHECK **3A: Geometric series**

1 For the following geometric series, find

   i the seventh term   **ii** the sum of the first seven terms   **iii** the sum to infinity (if possible):

   a $2 + 10 + 50 + \cdots$ **b** $7 + \tfrac{7}{2} + \tfrac{7}{4} + \cdots$ **c** $1 - 2 + 4 - 8 + \cdots$

2 Find the sum of the first ten terms of the geometric series $-2 - 6 - 18 - \cdots$

3 A property was valued at £70 000 at the start of 2001. If the projected increase in value is 3% per year, find the projected value of the property at the start of 2020. Give your answer to the nearest £1000.

4 A geometric series has first term a and common ratio r, where $r > 0$. The third term is $\frac{5}{2}$ and the seventh term is $\frac{5}{512}$.

   a Find the values of a and r.

   b Find the sum to infinity of the series.

5 The sum to infinity of a geometric series is $\frac{3}{4}$ and the sum of the first two terms is $\frac{2}{3}$. The common ratio of the series is negative.

   a Find the common ratio.

   b Find the *exact* difference between the sum of the first five terms and the sum to infinity.

6 Evaluate   **a** $\displaystyle\sum_{r=1}^{9} 2(4^r)$ **b** $\displaystyle\sum_{r=1}^{\infty} \left(\tfrac{3}{4}\right)^r$.

7 Evaluate $\displaystyle\sum_{r=1}^{6} (7^r + 1)$.

8 On her first birthday, Belinda is given £1. In each subsequent year, she is given double the amount that she received in the previous year, so that she receives £2 on her second birthday, £4 on her third birthday and so on.

   a How much does she receive on her tenth birthday?

   b How much, in total, has she received when she is ten?

9 A geometric series has first term 2 and second term $2\sqrt{2}$.

   a Find the seventh term.

   b The sum of the first five terms is $p + q\sqrt{2}$. Find the values of p and q.

10 Evaluate $\displaystyle\sum_{r=1}^{\infty} 2^{-r}$.

SKILLS CHECK 3A EXTRA is on the CD

3.2 Binomial expansions

The binomial expansion of $(1 + x)^n$ for positive integer n; the notations $n!$ and $\binom{n}{r}$.

You can multiply brackets containing two terms, with a low power outside the bracket, such as $(1 + 3x)^2$ or $(2x - 3y)^3$, by simply multiplying the brackets out.

When the power of the bracket is higher, you can use the **binomial expansion formula**.

Notice the pattern in these expansions:

$$(a + b)^0 = \qquad\qquad 1$$
$$(a + b)^1 = \qquad\qquad a + b$$
$$(a + b)^2 = \qquad\qquad a^2 + 2ab + b^2$$
$$(a + b)^3 = \qquad\quad a^3 + 3a^2b + 3ab^2 + b^3$$
$$(a + b)^4 = \qquad a^4 + 4a^3b + 4a^2b^2 + 4ab^3 + b^4$$
$$(a + b)^5 = \quad a^5 + 5a^4b + 10a^3b^2 + 10a^2b^3 + 5ab^4 + b^5$$

Note:
These are called binomial expansions because they contain two terms in the bracket.

The coefficients of the terms can be found using **Pascal's triangle**, in which a number is formed by adding the two numbers immediately above it.

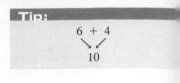

Another way of finding the coefficients is to use $\binom{n}{r}$, where

$$\binom{n}{r} = \frac{n!}{r!(n-r)!}$$

$\binom{n}{r}$ is sometimes referred to as ^nC_r and its value can be found directly from a calculator, using the $\boxed{nCr}$ button.

Tip:
$a = 1a$, so the coefficient of a is 1.

Note:
$n! = n(n-1)(n-2)$
$\dots \times 3 \times 2 \times 1$

Binomial expansion formula for $(a + b)^n$

In general, for positive integer values of n,

$$(a+b)^n = \binom{n}{0}a^n b^0 + \binom{n}{1}a^{n-1}b^1 + \dots + \binom{n}{r}a^{n-r}b^r + \dots + \binom{n}{n}a^0 b^n$$

$$= a^n + \binom{n}{1}a^{n-1}b^1 + \binom{n}{2}a^{n-2}b^2 + \dots + \binom{n}{r}a^{n-r}b^r + \dots + b^n$$

Note:
(i) The a-term descends in power, (ii) the b-terms ascends in power, (iii) the powers of a and b add up to the power in the expansion.

Note:
In the examination, this formula will be given.

Example 3.8 Using the binomial formula, find the expansion of $(2x + 3x)^4$ in ascending powers of x.

Step 1: Compare with $(a + b)^n$ and identify the unknown variables.

Comparing with $(a + b)^n$, you have $a = 2$, $b = 3x$, $n = 4$.

Step 2: Substitute into the binomial expansion formula.

$$(2 + 3x)^4 = 2^4 + \binom{4}{1}2^3(3x) + \binom{4}{2}2^2(3x)^2 + \binom{4}{3}2(3x)^3 + (3x)^4$$

Step 3: Simplify the terms.

$$= 16 + 96x + 216x^2 + 216x^3 + 81x^4$$

Tip:
$a = a^1$
$\binom{n}{0}a^n b^0 = a^n$, $\binom{n}{n}a^0 b^n = b^n$

Example 3.9 Given that $(4 - 5y)^9 = A + By + Cy^2 + Dy^3 + \dots$, find A, B, C and D.

Step 1: Compare with $(a + b)^n$ and identify the unknown variables.

Comparing with $(a + b)^n$, you have $a = 4$, $b = -5y$, $n = 9$.

Step 2: Substitute into the binomial expansion formula.

$$(4 - 5y)^9 = 4^9 + \binom{9}{1}4^8(-5y) + \binom{9}{2}4^7(-5y)^2 + \binom{9}{3}4^6(-5y)^3 + \dots$$

Step 3: Simplify the terms.

$$= 262\,144 - 2\,949\,120y + 14\,745\,600y^2 - 43\,008\,000y^3 + \dots$$

$A = 262\,144$, $B = -2\,949\,120$, $C = 14\,745\,600$, $D = 43\,008\,000$.

Tip:
This is the same as calculating the first four terms in the expansion.

You can easily find individual terms of the expansion $(a + b)^n$ without having to expand the whole series, by using the fact that the bottom number of $\binom{n}{r}$ is the same as the power of the b-term and the sum of the powers of a and b is always n.

Example 3.10 Find the coefficient of x^4 in the expansion of $(2 + \frac{1}{2}x)^5$.

Step 1: Identify and calculate the term required.
The term required is $\binom{5}{4}2^1(\frac{1}{2}x)^4 = 5 \times 2 \times \frac{1}{16}x^4 = \frac{5}{8}x^4$.

Step 2: State the coefficient.
The coefficient of x^4 is $\frac{5}{8}$.

> **Tip:**
> The fifth term is the term containing x^4.

Properties of $n!$ and $\binom{n}{r}$

1 $n! = n(n - 1)(n - 2)\ldots 4.3.2.1$
   $= n(n - 1)!$

So, for example,
$5! = 5 \times 4 \times 3 \times 2 \times 1 = 120$
   $= 5.4!$

> **Note:**
> The dots here are not decimals but are short for multiplication signs.

2 $\binom{n}{0} = 1$

because $\binom{n}{0} = \dfrac{n!}{0!(n - 0)!} = \dfrac{n!}{n!} = 1$

So, for example, $\binom{5}{0} = 1, \binom{7}{0} = 1$.

3 $\binom{n}{1} = n$

because $\binom{n}{1} = \dfrac{n!}{1!(n - 1)!} = \dfrac{n(n - 1)!}{(n - 1)!} = n$

So, for example, $\binom{5}{1} = 5$ and $\binom{25}{1} = 25$.

4 $\binom{n}{2} = \dfrac{n(n - 1)}{2}$

because $\binom{n}{2} = \dfrac{n!}{2!(n - 2)!} = \dfrac{n(n - 1)(n - 2)!}{2(n - 2)!} = \dfrac{n(n - 1)}{2}$

> **Note:**
> $10! = 10.9.8!$

> **Tip:**
> These are properties of $\binom{n}{r}$ that you should know how to prove for yourself.

5 $\binom{n}{3} = \dfrac{n(n - 1)(n - 2)}{6}$

because $\binom{n}{3} = \dfrac{n!}{3!(n - 3)!} = \dfrac{n(n - 1)(n - 2)(n - 3)!}{6(n - 3)!} = \dfrac{n(n - 1)(n - 2)}{6}$

Binomial expansion formula for $(1 + x)^n$

If you substitute $a = 1$ and $b = x$ into the original expansion, you get a special case of the binomial expansion that is often used to make calculation easier:

> **Tip:**
> Remember that $1^m = 1$ for any integer m.

$$(1 + x)^n = 1 + \binom{n}{1}x + \binom{n}{2}x^2 + \cdots + x^n$$

$$= 1 + nx + \frac{n(n - 1)}{2 \times 1}x^2 + \cdots + x^n$$

Example 3.11   Show that $(1 + \sqrt{3})^4 + (1 - \sqrt{3})^4$ is an integer. State the value of this integer.

Step 1: Compare with $(1 + x)^n$ and identify the unknown variables.

For the first bracket, $x = \sqrt{3}$, $n = 4$.

For the second bracket, $x = -\sqrt{3}$, $n = 4$.

Step 2: Substitute into the binomial expansion formula.

$$(1 + \sqrt{3})^4 = 1 + 4(\sqrt{3}) + \frac{4 \times 3}{2 \times 1}(\sqrt{3})^2 + \frac{4 \times 3 \times 2}{3 \times 2 \times 1}(\sqrt{3})^3 + (\sqrt{3})^4$$

$$= 1 + 4\sqrt{3} + 18 + 12\sqrt{3} + 9 = 28 + 16\sqrt{3}$$

Similarly,

$$(1 - \sqrt{3})^4 = 1 - 4\sqrt{3} + 18 - 12\sqrt{3} + 9 = 28 - 16\sqrt{3}$$

Step 3: Simplify the terms.

And so, $(1 + \sqrt{3})^4 + (1 - \sqrt{3})^4 = 28 + 16\sqrt{3} + 28 - 16\sqrt{3} = 56$.

This is an integer with value 56.

> **Tip:**
> We only need to expand $(1 + \sqrt{3})^4$ because the expansion of $(1 - \sqrt{3})^4$ is similar with every term containing an odd power of $\sqrt{3}$ having a minus sign instead of a plus sign.

Example 3.12   Use the first four terms in the expansion of $(1 - 2x)^8$ to find an approximate value for $(0.98)^8$.

Step 1: Compare with $(1 + x)^n$ and identify the unknown variables.

Expand $(1 - 2x)^8$ in ascending powers of x as far as the term in x^3. $n = 8$, and the x-term is $-2x$.

Step 2: Substitute into the binomial expansion formula.

$$(1 - 2x)^8 = 1 + 8(-2x) + \frac{8 \times 7}{2 \times 1}(-2x)^2 + \frac{8 \times 7 \times 6}{3 \times 2 \times 1}(-2x)^3 + \cdots$$

Step 3: Simplify the terms.

$$= 1 - 16x + 112x^2 - 448x^3 + \cdots$$

Step 4: Compare the value with the terms in the bracket of the expansion and solve for x.

We require $(0.98)^8$.

Comparing this with $(1 - 2x)^8$ gives

$0.98 = 1 - 2x \Rightarrow x = 0.01$

Step 5: Substitute this x-value into the expansion.

Substitute $x = 0.01$ into the expansion:

$$(0.98)^8 = 1 - 16(0.01) + 112(0.01)^2 - 448(0.01)^3 + \cdots$$

$$= 0.850\,752$$

$$= 0.851 \text{ (3 d.p.)}$$

> **Tip:**
> The three dots in the expansion are there to show that there are other terms.

Example 3.13   In the expansion of $(1 + kx)^n$, where k and n are positive integers, the coefficient of x is 15 and the coefficient of x^2 is 90. Find n and k. Hence find the term in x^3.

Step 1: Compare with $(1 + x)^n$ and identify the unknown variables.

The power is n and the x-term is kx.

Step 2: Substitute into the binomial expansion formula.

$$(1 + kx)^n = 1 + nkx + \frac{n(n - 1)}{2}k^2x^2 + \frac{n(n - 1)(n - 2)}{3 \times 2 \times 1}k^3x^3 + \cdots$$

Step 3: Simplify the terms.

Coefficient of x: $nk = 15$ ①

Step 4: Compare the terms and solve for the unknowns.

Coefficient of x^2: $\dfrac{n(n - 1)}{2}k^2 = 90$ ②

Now solve simultaneously:

From ①: $k = \dfrac{15}{n}$ ③

Substitute into ②: $\dfrac{n(n - 1)}{2}\left(\dfrac{15}{n}\right)^2 = 90$

Simplifying: $\dfrac{225n(n - 1)}{2n^2} = 90$

$\Rightarrow 225(n - 1) = 180n$

$\Rightarrow 225n - 225 = 180n \Rightarrow n = 5$

> **Tip:**
> Cancel the n and then multiply by $2n$.

Substitute into ③: $$k = \frac{15}{n} = 3$$

Hence n is 5 and k is 3.

The term in x^3 is $\dfrac{5 \times 4 \times 3}{3 \times 2 \times 1}(3x)^3 = 270x^3$.

3B: Binomial expansions

1. The polynomial f(x) is given by $(2 - 3x)^4$. Find the binomial expansion of f(x), simplifying your terms.

2. Simplifying your terms, find the first four terms in the expansion of $(1 + 4y)^7$, in ascending powers of y.

3. a Expand $(3 - 2x)^5$ in ascending powers of x up to the term in x^2.

   b Find the values of A, B and C, where $(5 + 2x)(3 - 2x)^5 = A + Bx + Cx^2 + \cdots$

4. a Expand $(1 + 2x)^6$ in ascending powers of x up to the term in x^3.

   b Using your expansion, find an approximation for $(1.02)^6$, correct to four decimal places. You must write down sufficient working to show how you obtained your answer.

5. Using the first four terms, in ascending powers of x, of the expansion of $(1 - 4x)^7$, find an approximate value for $(0.996)^7$, to a suitable degree of approximation. You must write down sufficient working to show how you obtained your answer.

6. Expand and simplify $(\sqrt{2} + \sqrt{3})^4 - (\sqrt{2} - \sqrt{3})^4$, leaving your answer in the form $a\sqrt{6}$, where a is a positive integer.

7. The coefficient of x^2 is $\frac{3}{8}$ in the expansion of $\left(1 + \dfrac{x}{n}\right)^n$. Find the value of n.

8. It is given that $(1 + kx)^n = 1 - 4x + 7x^2 + \cdots$

   a Find n and k.

   b Hence find the term in x^3.

9. a Expand $(1 + ax)^6$ in ascending powers of x up to and including the term in x^2.

   b In the expansion $(1 + bx)(1 + ax)^6$, the coefficients of x and x^2 are 20 and 171 respectively. Find a and b, given that they are integers.

10. a Write down the first four terms, in ascending powers of x, in the expansion of $(1 - 3x)^5$.

    b Find the coefficient of x^3 in the expansion of $(1 + x)(1 - 3x)^5$.

1 The third and fourth terms of a geometric series are 6.4 and 5.12 respectively. Find

   a the common ratio of the series,

   b the first term of the series,

   c the sum to infinity of the series.

   d Calculate the difference between the sum to infinity of the series and the sum of the first 25 terms of the series. [Edexcel Jan 2001]

2 The second term of a geometric series is 80 and the fifth term of the series is 5.12.

   a Find the common ratio and the first term of the series.

   b Find the sum to infinity of the series, giving your answer as an exact fraction.

   c Find the difference between the sum to infinity of the series and the sum of the first 14 terms of the series, giving your answer in the form $a \times 10^n$, where $1 \leqslant a < 10$ and n is an integer. [Edexcel Specimen Paper]

3 **a** Expand $(2 + \frac{1}{4}x)^9$ in ascending powers of x as far as the term in x^3, simplifying each term.

   b Use your series, together with a suitable value of x, to calculate an estimate of $(2.025)^9$. [Edexcel Mock Paper]

4 **a** Write down the first 4 terms of the binomial expansion, in ascending powers of x, of $(1 + ax)^n$, $n > 2$.

   Given that, in this expansion, the coefficient of x is 8 and the coefficient of x^2 is 30,

   b calculate the value of n and the value of a,

   c find the coefficient of x^3. [Edexcel Nov 2003]

5 Initially the number of fish in a lake is 500 000. The population is then modelled by the recurrence relation

$$u_{n+1} = 1.05u_n - d, \quad u_0 = 500\,000.$$

In this relation u_n is the number of fish in the lake after n years and d is the number of fish which are caught each year.

Given that $d = 15\,000$,

   a calculate u_1, u_2 and u_3 and comment briefly on your results.

Given that $d = 100\,000$,

   b show that the population of fish dies out during the sixth year.

   c Find the value of d which would leave the population each year unchanged. [Edexcel Jan 2002]

6 $f(x) = \left(1 + \dfrac{x}{k}\right)^n$, $k, n \in \mathbb{N}$, $n > 2$.

Given that the coefficient of x^3 is twice the coefficient of x^2 in the binomial expansion of $f(x)$,

a prove that $n = 6k + 2$.

Given also that the coefficients of x^4 and x^5 are equal and non-zero,

b form another equation in n and k and hence show that $k = 2$ and $n = 14$.

Using these value of k and n,

c expand $f(x)$ in ascending powers of x, up to and including the term in x^5. Give each coefficient as an exact fraction in its lowest terms.

[Edexcel Jan 2002]

7 The first three terms in the expansion, in ascending powers of x, of $(1 + px)^n$, are $1 - 18x + 36p^2x^2$. Given that n is a positive integer, find the value of n and the value of p.

[Edexcel Jan 2003]

8 The expansion of $(2 - px)^6$ in ascending powers of x, as far as the term in x^2, is

$$64 + Ax + 135x^2.$$

Given that $p > 0$, find the value of p and the value of A. [Edexcel June 2003]

9 a Write down the first four terms of the binomial expansion, in ascending powers of x, of $(1 + 3x)^n$, where $n > 2$.

Given that the coefficient of x^3 in this expansion is ten times the coefficient of x^2,

b find the value of n,

c find the coefficient of x^4 in the expansion. [Edexcel June 2002]

4 Trigonometry

4.1 Sine and cosine rules

The sine and cosine rules.

The sine and cosine rules are used to find lengths and angles in a triangle.

To apply them, you need to label your triangle as follows:

Label the vertices with upper case letters, for example *A*, *B* and *C*. Then label the side opposite each vertex with the corresponding lower case letter, so side *a* is opposite angle *A*, side *b* is opposite angle *B* and side *c* is opposite angle *C*.

Note:
The triangle can be any size and shape.

Note:
A **vertex** is where two lines meet. The plural is **vertices**.

The sine rule

To find a **length**, use

$$\frac{a}{\sin A} = \frac{b}{\sin B} = \frac{c}{\sin C}$$

Format (1)

To find an **angle**, use

$$\frac{\sin A}{a} = \frac{\sin B}{b} = \frac{\sin C}{c}$$

Format (2)

Tip:
You can use the sine rule to find:
• a side when you know the angles in the triangle and a side
• an angle when you know two sides and the angle opposite one of them.

Tip:
To proceed, you must know one complete ratio.

Note:
You must learn the sine rule.

Example 4.1 In triangle *ABC*, *AC* is 3.2 cm, angle *ABC* is 35° and angle *BCA* is 82°. Find *AB*, giving your answer to the nearest mm.

Step 1: Draw a carefully labelled sketch and include all known measures.

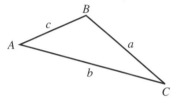

Tip:
Drawing a sketch will help you to see the situation and plan a strategy.

Step 2: Write down the sine rule in format (1) for finding a length.

Using the sine rule:

$$\frac{a}{\sin A} = \frac{b}{\sin B} = \frac{c}{\sin C}$$

Step 3: Substitute known values.

$$\frac{a}{\sin A} = \frac{3.2}{\sin 35°} = \frac{c}{\sin 82°}$$

Step 4: Choose the two relevant ratios and solve the equation.

$$\frac{c}{\sin 82°} = \frac{3.2}{\sin 35°}$$

$$c = \frac{3.2 \times \sin 82°}{\sin 35°}$$

$$= 5.5247...$$

So *AB* = 5.5 cm (to nearest mm)

Note:
You know length *b* and angle *B*. You are not asked anything about *a* and *A* so ignore the ratio involving them.

Note:
To give the answer to the nearest mm, you need to correct your value in cm to one decimal place.

Example 4.2 Find θ, giving your answer to the nearest degree.

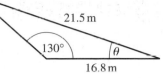

Step 1: Label the sketch carefully.

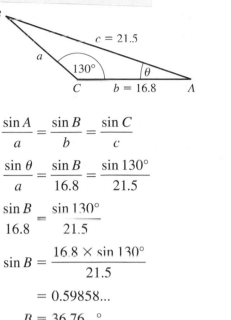

Note:

θ is angle A.

Step 2: Write down the sine rule in format (2) for finding an angle.

$$\frac{\sin A}{a} = \frac{\sin B}{b} = \frac{\sin C}{c}$$

Step 3: Substitute known values.

$$\frac{\sin \theta}{a} = \frac{\sin B}{16.8} = \frac{\sin 130°}{21.5}$$

Step 4: Choose the two relevant ratios and solve the equation.

$$\frac{\sin B}{16.8} = \frac{\sin 130°}{21.5}$$

$$\sin B = \frac{16.8 \times \sin 130°}{21.5}$$

$$= 0.59858...$$

$$B = 36.76...°$$

$$\Rightarrow \quad \theta = 180° - (130° + 36.76...°)$$

$$= 13.23...°$$

$$= 13° \text{ (nearest degree.)}$$

Note:

As you do not know a, you cannot use $\dfrac{\sin \theta}{a}$, but you *can* use the middle ratio to find angle B. You can then calculate θ.

Note:

$B = \sin^{-1}(0.59858...)$

Tip:

The sum of the angles in a triangle is 180°.

The cosine rule

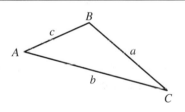

To find a **length**, for example a, use this formula for a^2:

$$a^2 = b^2 + c^2 - 2bc \cos A \qquad \qquad Format\ (1)$$

Then square root to calculate a.

To find b, use

$$b^2 = a^2 + c^2 - 2ac \cos B$$

To find c, use

$$c^2 = a^2 + b^2 - 2ab \cos C$$

To find an **angle**, re-arrange the formulae as follows:

$$\cos A = \frac{b^2 + c^2 - a^2}{2bc} \qquad \qquad Format\ (2)$$

$$\cos B = \frac{a^2 + c^2 - b^2}{2ac}$$

$$\cos C = \frac{a^2 + b^2 - c^2}{2ab}$$

Tip:

You can use the cosine rule to find:

- the third side when you know two sides and the angle between them
- an angle when you know the three sides.

Note:

You will be given format (1) in the examination.

Make sure you can rearrange it to find an angle. Remember the patterns.

29

Example 4.3 In triangle ABC, $AC = 6.2$ cm, $BC = 8.3$ cm and angle $ACB = 42°$.
Calculate AB, giving your answer correct to three significant figures.

Step 1: Draw a carefully labelled sketch and include all known measures.

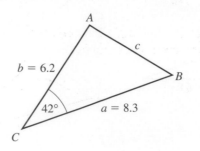

Tip:
The cosine rule is appropriate as you know two sides and the angle between them.

Step 2: Use the cosine rule to find the missing length.

By the cosine rule

$$c^2 = a^2 + b^2 - 2ab \cos C$$

$$= 8.3^2 + 6.2^2 - 2 \times 8.3 \times 6.2 \times \cos 42°$$

$$= 30.84\ldots$$

$$c = \sqrt{30.84\ldots} = 5.553\ldots$$

$$AB = 5.55 \text{ cm (3 s.f.)}$$

Tip:
Find c^2 in one stage on your calculator; do not press $=$ until you have entered cos 42.

Example 4.4 A triangle has sides of length 6 cm, 8 cm and 12 cm. Find the size of the largest angle in the triangle, giving your answer to the nearest $0.1°$.

Step 1: Draw a carefully labelled sketch and include all known measures.

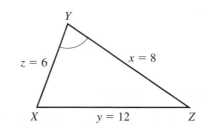

Tip:
The largest angle is opposite the longest side.

Step 2: Use the cosine rule to find the missing angle.

$$\cos Y = \frac{x^2 + z^2 - y^2}{2xz}$$

$$= \frac{8^2 + 6^2 - 12^2}{2 \times 8 \times 6}$$

$$= -0.458\ldots$$

$$Y = 117.3° \text{ (nearest } 0.1°)$$

Tip:
Giving your answer to the nearest $0.1°$ is the same as approximating to one decimal place.

Calculator note:

There are several ways of entering the calculation. Make sure that the method you use is a correct one.

Here is an example:

Tip:
The denominator must be enclosed in brackets here.

The area of a triangle in the form $\frac{1}{2}ab \sin C$.

When you know **two sides** and the **angle between them**, you can use this formula to calculate the **area** of a triangle.

$$\text{Area} = \tfrac{1}{2}ab \sin C$$

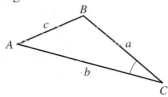

You need to rearrange the formula if you know angle A or angle B as follows:

$$\text{Area} = \tfrac{1}{2}bc \sin A$$

$$\text{Area} = \tfrac{1}{2}ac \sin B$$

Example 4.5 In triangle PQR, $PQ = 4.2$ cm, $QR = 6.3$ cm and angle $PQR = 130°$. Calculate the area of the triangle, giving your answer to two significant figures.

Step 1: Draw a carefully labelled sketch and include all known measures.

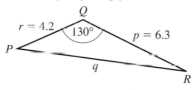

Step 2: Use the area formula.

$$\text{Area} = \tfrac{1}{2}pr \sin Q$$
$$= \tfrac{1}{2} \times 6.3 \times 4.2 \times \sin 130°$$
$$= 10.13\ldots$$
$$= 10 \text{ cm}^2 \text{ (2 s.f.)}$$

Example 4.6 A triangle has sides of length 5 cm, 8 cm and 9 cm.

Calculate the area of the triangle, giving your answer correct to two significant figures.

Step 1: Draw a carefully labelled sketch and include all known measures.

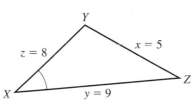

Step 2: Use the cosine rule to find an angle.

Calculate an angle, using the cosine rule.

$$\cos X = \frac{y^2 + z^2 - x^2}{2yz}$$
$$= \frac{9^2 + 8^2 - 5^2}{2 \times 9 \times 8}$$
$$= 0.833\ldots$$
$$X = 33.557\ldots°$$

Tip:
It does not matter which angle you calculate.

Step 3: Use the area formula.

$$\text{Area} = \tfrac{1}{2}yz \sin X$$
$$= \tfrac{1}{2} \times 9 \times 8 \times \sin 33.557..°$$
$$= 19.89\ldots$$
$$= 20 \text{ cm}^2 \text{ (2 s.f.)}$$

Tip:
Do not approximate here but use the full display on the calculator in the next calculation.

Give answers to three significant figures unless requested otherwise.

1 Calculate *x*, correct to the nearest mm.

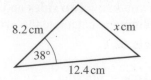

8.2 cm

x cm

38°

12.4 cm

2 Calculate *y*.

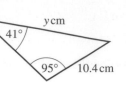

y cm

41°

95° 10.4 cm

3 a Calculate angle *ABC*.

b Calculate the area of triangle *ABC*.

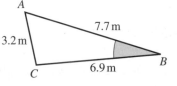

A

7.7 m

3.2 m

B

C 6.9 m

4 In triangle *QRP*, *QR* = 4 mm, *RP* = 5.5 mm, angle *QPR* = 35°.

a Calculate angle *PQR*, given that it is acute.

b Calculate angle *QRP*.

Q

4 mm

35° *P*

5.5 mm

R

5 In triangle *ABC*, angle *BAC* = 15°, angle *ABC* = 140° and *AC* = 20.5 cm.

a Calculate length *BC*. **b** Calculate the area of triangle *ABC*.

6 A triangle has two equal sides of length 6 cm and one of the angles in the triangle is 40°.

a Sketch two possible triangles.

b For each triangle, calculate
i the area of the triangle,
ii the length of the third side.

7 Calculate angle *QRS*.

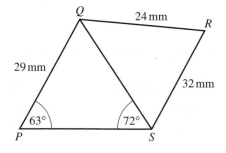

Q 24 mm *R*

29 mm

32 mm

63° 72°

P *S*

8 A circle, centre *O*, has radius 12 cm. Angle *OPQ* = 37°. Calculate

a the length of the chord *PQ*,

b the area of triangle *OPQ*.

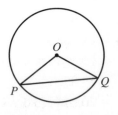

O

P *Q*

9 *ABCD* is a field in the shape of a quadrilateral.
AB = 25 m, *AD* = 15 m, *DC* = 17 m.
Angle *BAD* = 45°, angle *BCD* = 62°. Calculate

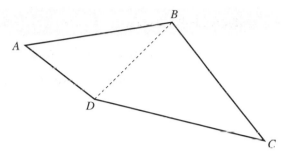

  a the length of the diagonal *BD*,

  b angle *DBC*,

  c angle *BDC*,

  d the area of the field.

10 In triangle *ABC*, *AB* = 8 cm and *AC* = 12 cm.

  a The area of the triangle is 24 cm². Given that angle *BAC* is acute, calculate angle *BAC* and length *BC*.

  b Show that when angle *BAC* is 150°, the area of triangle *ABC* is also 24 cm² and calculate the length *BC* in this case.

  c With the aid of diagrams, comment on your answers to parts **a** and **b**.

SKILLS CHECK **4A EXTRA is on the CD**

4.3 Degrees and radians

Degree and radian measure.

When the radius *OP* turns through an angle θ about *O*, the **sector** *POQ* is formed, with **arc length** *PQ*.

When the length of the arc is the same as the radius, the angle is 1 **radian** (1ᶜ).

Note:
The symbol ᶜ stands for circular measure.

Converting between radians and degrees

Make sure that you learn the following:
  π radians = 180°

To convert radians to degrees, multiply by $\dfrac{180}{\pi}$.

To convert degrees to radians, multiply by $\dfrac{\pi}{180}$.

Note:
In one complete turn of 360°, the arc length is $2\pi r$ (2π lots of *r*), so the angle is 2π lots of 1 radian, that is 2π radians. This means that 2π radians = 360°.

Example 4.7 Write 40° in radians as a multiple of π.

Step 1: Multiply by $\dfrac{\pi}{180}$
  $40° = 40 \times \dfrac{\pi}{180}$ radians $= \dfrac{2}{9}\pi$ radians

and simplify, leaving π in your answer.

Tip:
When an angle is given as a multiple of π, the word radians or the symbol ᶜ is usually omitted.

You will find it helpful to learn these common conversions:

Radians	Degrees
$\frac{1}{6}\pi$	30°
$\frac{1}{4}\pi$	45°
$\frac{1}{3}\pi$	60°
$\frac{1}{2}\pi$	90°
π	180°
2π	360°
1ᶜ	57° (nearest degree)
0.017ᶜ (3 d.p.)	1°

Tip:
If you forget, use 180° = π radians and work them out yourself.

Arc length, area of sector.

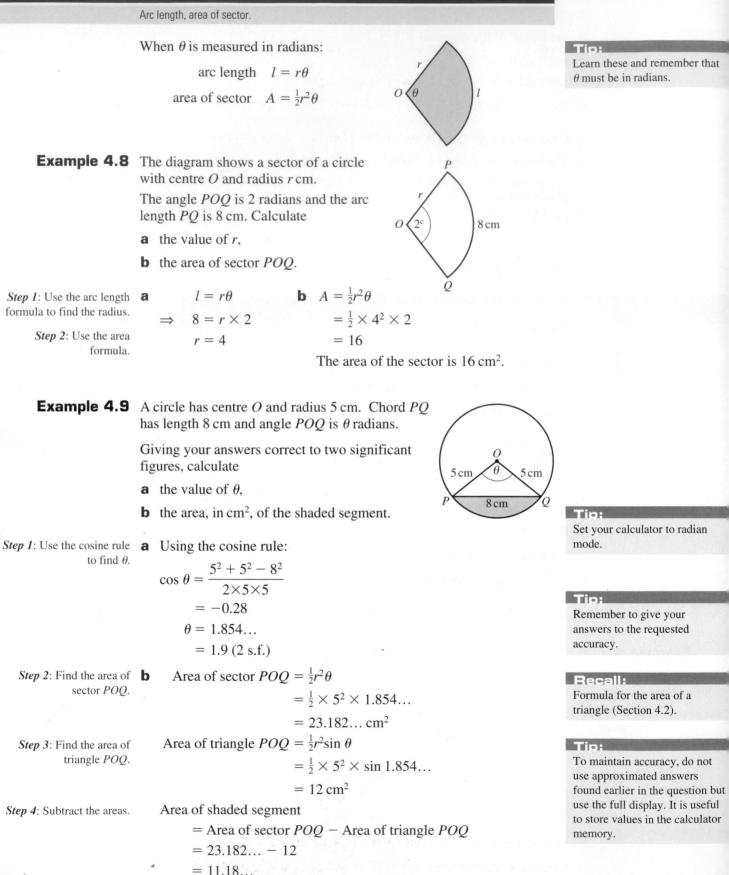

When θ is measured in radians:

$$\text{arc length} \quad l = r\theta$$

$$\text{area of sector} \quad A = \tfrac{1}{2}r^2\theta$$

Example 4.8 The diagram shows a sector of a circle with centre O and radius r cm.

The angle POQ is 2 radians and the arc length PQ is 8 cm. Calculate

a the value of r,

b the area of sector POQ.

Step 1: Use the arc length formula to find the radius.

a $\quad l = r\theta$

$\Rightarrow \quad 8 = r \times 2$

$r = 4$

Step 2: Use the area formula.

b $\quad A = \tfrac{1}{2}r^2\theta$

$= \tfrac{1}{2} \times 4^2 \times 2$

$= 16$

The area of the sector is 16 cm^2.

Example 4.9 A circle has centre O and radius 5 cm. Chord PQ has length 8 cm and angle POQ is θ radians.

Giving your answers correct to two significant figures, calculate

a the value of θ,

b the area, in cm^2, of the shaded segment.

Step 1: Use the cosine rule to find θ.

a Using the cosine rule:

$$\cos \theta = \frac{5^2 + 5^2 - 8^2}{2 \times 5 \times 5}$$

$$= -0.28$$

$$\theta = 1.854\ldots$$

$$= 1.9 \text{ (2 s.f.)}$$

Step 2: Find the area of sector POQ.

b $\quad$ Area of sector $POQ = \tfrac{1}{2}r^2\theta$

$$= \tfrac{1}{2} \times 5^2 \times 1.854\ldots$$

$$= 23.182\ldots \text{ cm}^2$$

Step 3: Find the area of triangle POQ.

Area of triangle $POQ = \tfrac{1}{2}r^2\sin \theta$

$$= \tfrac{1}{2} \times 5^2 \times \sin 1.854\ldots$$

$$= 12 \text{ cm}^2$$

Step 4: Subtract the areas.

Area of shaded segment

$= $ Area of sector $POQ - $ Area of triangle POQ

$= 23.182\ldots - 12$

$= 11.18\ldots$

$= 11 \text{ cm}^2$ (2 s.f.)

Give answers to three significant figures unless requested otherwise.

1 Convert   **a**   280° to radians   **b**   1.5 radians to degrees.

2 Convert the following angles in radians to degrees.

   a $\frac{2}{3}\pi$ **b** $\frac{3}{4}\pi$ **c** $\frac{3}{2}\pi$ **d** $\frac{7}{12}\pi$

3 Convert these angles to radians, giving each angle in terms of π.

   a   45°   **b**   150°   **c**   330°   **d**   240°

4 *POQ* is a sector of a circle, centre *O*, radius 5 cm.
Angle *POQ* is 0.6 radians. Calculate

   a   the length of the arc *PQ*,   **b**   the perimeter of the sector,

   c   the area of triangle *POQ*,   **d**   the area of sector *POQ*.

5 *AOB* is a sector of a circle, centre *O*, radius 10.4 cm. The arc length *AB* is 12.48 cm.

   a   Find angle *AOB*.   **b**   Find the area of sector *AOB*.

6 A sector of a circle has area 27 cm² and radius 6 cm.

   a   Calculate the angle of the sector, in radians.

   b   Calculate the perimeter of the sector.

7 A circle, with centre *O*, has radius 8 cm.
A chord intersects the circle at *P* and *Q* and
angle *POQ* is θ radians, where θ is acute.

   The area of triangle *POQ* is 24 cm². Find

   a   the value of θ,

   b   the area of sector *POQ*,

   c   the area of the shaded segment.

8 *PQ* is an arc of a circle, centre *A*, radius 10 cm. *BC* is an arc of a circle, centre *A*, radius 7 cm.
The size of angle *PAQ* is θ radians.

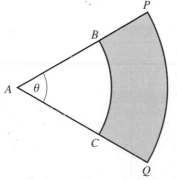

   a   Find, in terms of θ, an expression for the perimeter of *BPQC*.

   b   Given that the perimeter of *BPQC* is 14.5 cm, show that θ is 0.5.

   c   Find, in cm², the area of *BPQC*.

9 In triangle ABC, $AB = 9\,\text{cm}$, $AC = 6\,\text{cm}$ and angle $BAC = \frac{1}{6}\pi$ radians.
A circle, centre A, radius 2 cm, intersects the triangle at P and Q.
Calculate

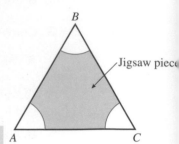

   a length BC,   **b** arc length QP,

   c the perimeter of the shaded region $QBCP$,   **d** the area of triangle ABC,

   e the area of sector AQP,   **f** the area of the shaded region $QBCP$.

10 A jigsaw piece is made from an equilateral triangle
ABC with sides of length 2 cm.
A sector of a circle, radius 0.5 cm, is cut away from each vertex.

   a Find the perimeter of the jigsaw piece.

   b Find the area of the jigsaw piece.

Jigsaw piece

SKILLS CHECK **4B EXTRA** is on the CD

4.5 Trigonometric functions

Sine, cosine and tangent functions. Their graphs, symmetries and periodicity.

You will need to be able to recall the main features of the graphs of
$y = \sin x$, $y = \cos x$ and $y = \tan x$. Make sure that you can sketch
them for x in degrees or radians.

Remember that $180° = \pi$ radians.

Note:
sin is shorthand for sine, cos for
cosine and tan for tangent.

Tip:
In module C2 you may use a
graphical calculator to check.

$y = \sin x$

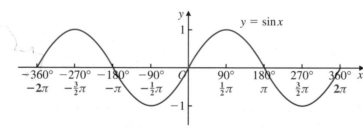

Note:
In degrees:
$\sin(x + 360°) = \sin x$.
In radians:
$\sin(x + 2\pi) = \sin x$.

The minimum value of $\sin x$ is -1 and the maximum value is 1.
The graph is periodic, repeating every $360°$ (2π radians).
The vertical line through every vertex (turning point) is an axis of
symmetry.

$y = \cos x$

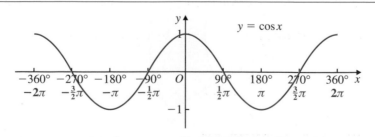

Recall:
$y = \cos x$ is a translation of
$y = \sin x$ by $90°$ $\left(\frac{1}{2}\pi\right)$ to the left,
i.e. $\cos x = \sin(x + 90°)$,
or $\sin\left(x + \frac{1}{2}\pi\right)$ in radians.

The minimum value of $\cos x$ is -1 and the maximum value is 1.
The graph is periodic, repeating every $360°$ (2π radians).
The vertical line through every vertex (turning point) is an axis of
symmetry, in particular the y-axis.

Note:
In degrees:
$\cos(x + 360°) = \cos x$.
In radians:
$\cos(x + 2\pi) = \cos x$.

$y = \tan x$

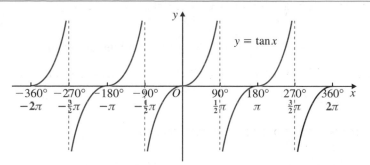

Note:
$\tan x$ takes every possible value in each $180°$ interval (π radians).

Notice that $\tan x$ can take any value.
The graph is periodic, repeating every $180°$ (π radians).
There are no lines of symmetry.

The graph has **asymptotes** at $x = \pm 90°$ ($\pm\frac{1}{2}\pi$), $x = \pm 270°$ ($\pm\frac{3}{2}\pi$), and so on.

Note:
In degrees:
$\tan (x + 180°) = \tan x$.
In radians:
$\tan(x + \pi) = \tan x$.

Note:
tan x is undefined at $+90°$, $\pm270°$, $\pm450°$, ...

Special angles

It is useful to recognise the sin, cos and tan of these special angles.

x	$\sin x$	$\cos x$	$\tan x$
$30°$ ($\frac{1}{6}\pi$)	$\dfrac{1}{2}$	$\dfrac{\sqrt{3}}{2}$	$\dfrac{1}{\sqrt{3}}$
$45°$ ($\frac{1}{4}\pi$)	$\dfrac{1}{\sqrt{2}}$	$\dfrac{1}{\sqrt{2}}$	1
$60°$ ($\frac{1}{3}\pi$)	$\dfrac{\sqrt{3}}{2}$	$\dfrac{1}{2}$	$\sqrt{3}$

Transformations of trigonometric graphs

You may be asked to sketch transformations of $y = \sin x$, $y = \cos x$ and $y = \tan x$, for example $y = \sin (x - 90°)$, $y = 2 \cos x$, $y = \tan 2x$, $y = \cos x + 1$.

In C1, you looked at transformations of functions. It would be worth reviewing these transformations, as they are now applied to the trigonometric functions.

Recall:
C1 Transformations
Section 1.13.

Example 4.10 **a** On the same set of axes, sketch the graphs of
  i $y = \sin x$ **ii** $y = \sin x + 1$
  in the interval $0° \leqslant x \leqslant 360°$.

b Hence, state the value of x at which the graph $y = \sin x + 1$ meets the x-axis.

Step 1: Identify the transformation.

Step 2: Sketch the graphs by applying the transformation.

a $y = \sin x + 1$ is a translation of $y = \sin x$ by 1 unit upwards.

Recall:
The graph of $y = f(x) + a$ is a translation of the graph of $y = f(x)$ by a units in the y-direction.

Recall:
If $a > 0$, the graph moves up.

b The graph of $y = \sin x + 1$ meets the x-axis at the point where $x = 270°$.

You should also be able to work in radians.

Example 4.11 Sketch the graph of $y = \cos\left(x - \frac{\pi}{2}\right)$, $0 \leq x \leq 2\pi$.

Step 1: Identify the transformation. $y = \cos\left(x - \frac{\pi}{2}\right)$ is a translation of $y = \cos x$ by $\frac{\pi}{2}$ radians to the right.

Step 2: Sketch the graphs by applying the transformation.

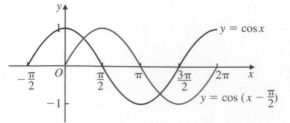

Recall:
The graph of $y = f(x + a)$ is a translation of the graph of $y = f(x)$ by $-a$ units in the x-direction.

Recall:
If $a < 0$, the graph moves right.

Note:
The new graph is in fact $y = \sin x$.

Example 4.12 Sketch the graph of $y = 3\sin x$ in the interval $0° \leq x \leq 360°$.

Step 1: Identify the transformation. $y = 3\sin x$ is a stretch of $y = \sin x$ by factor 3 in the y-direction.

Step 2: Sketch the graphs by applying the transformation.

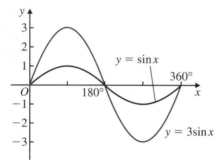

Recall:
The graph of $y = af(x)$ is a stretch of the graph of $y = f(x)$ by factor a in the y-direction.

Tip:
Stretch one point at a time; choose the main points to see where they are transformed to. Include axes intercepts and turning points.

Example 4.13 Sketch the graph of $y = \sin 3x$ for $0° \leq x \leq 360°$.

Step 1: Identify the transformation. $y = \sin 3x$ is a stretch of $y = \sin x$ by factor $\frac{1}{3}$ in the x-direction.

Step 2: Sketch the graphs by applying the transformation.

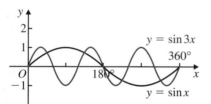

Recall:
The graph of $y = f(ax)$ is a stretch of the graph of $y = f(x)$ by factor $\dfrac{1}{a}$ in the x-direction.

Example 4.14 If $f(x) = \tan x$, $-90° \leq x \leq 90°$, sketch on the same set of axes

a $y = f(x)$ and $y = f(-x)$

b $y = f(x)$ and $y = -f(x)$

Step 1: Identify the transformation. **a** $y = f(-x)$ is a reflection of $f(x) = \tan x$ in the y-axis.

Step 2: Sketch the graphs by applying the transformation.

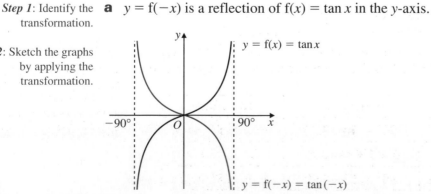

Recall:
The graph of $y = f(-x)$ is a reflection of the graph of $y = f(x)$ in the y-axis.

Tip:
Don't forget to include the asymptotes.

Step 1: Write down the transformation.

b $y = -f(x)$ is a reflection of $f(x) = \tan x$ in the x-axis.

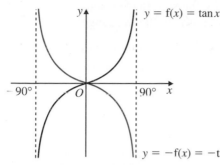

Step 2: Sketch the graphs by applying the transformation.

4.6 Trigonometric identities

Knowledge and use of $\tan \theta = \dfrac{\sin \theta}{\cos \theta}$ and $\sin^2\theta + \cos^2\theta = 1$.

The following **identities** are true for all values of θ.

$$\tan \theta \equiv \frac{\sin \theta}{\cos \theta}$$

$$\sin^2\theta + \cos^2\theta \equiv 1$$

Example 4.15 Find the exact value of $\tan \theta$, given that $\sin \theta = -\frac{5}{13}$ and $\cos \theta = \frac{12}{13}$.

Step 1: Use the appropriate trig identity and simplify if necessary.

$$\tan \theta = \frac{\sin \theta}{\cos \theta} = \frac{-\frac{5}{13}}{\frac{12}{13}} = -\frac{5}{12}$$

Example 4.16 Express $\dfrac{5 + 4\sin^2 \theta}{3 - 2\cos \theta}$ in the form $a + b\cos \theta$, where a and b are integers to be found.

Step 1: Write each expression in terms of $\cos \theta$.

$$5 + 4\sin^2 \theta = 5 + 4(1 - \cos^2 \theta)$$
$$= 5 + 4 - 4\cos^2 \theta$$
$$= 9 - 4\cos^2 \theta$$
$$= (3 - 2\cos \theta)(3 + 2\cos \theta)$$

Step 2: Simplify the given expression.

$$\frac{5 + 4\sin^2 \theta}{3 - 2\cos \theta} = \frac{(3 - 2\cos \theta)(3 + 2\cos \theta)}{3 - 2\cos \theta}$$

$$= 3 + 2\cos \theta$$

Step 3: Compare coefficients. Comparing with $a + b\cos \theta$ gives $a = 3$ and $b = 2$.

4.7 Trigonometric equations

Solution of simple trigonometric equations in a given interval of degrees or radians.

The simplest trigonometric equations are of the form $\sin x = c$, $\cos x = c$ and $\tan x = c$, where c is a number.

Your calculator will give you *one* solution to an equation of this type, the **principal value** (PV). This lies in a particular range, depending on the function.

	In degrees	In radians
For sine function	$-90° \leqslant PV \leqslant 90°$	$-\frac{1}{2}\pi \leqslant PV \leqslant \frac{1}{2}\pi$
For cosine function	$0 \leqslant PV \leqslant 180°$	$0 \leqslant PV \leqslant \pi$
For tangent function	$-90° \leqslant PV \leqslant 90°$	$-\frac{1}{2}\pi \leqslant PV \leqslant \frac{1}{2}\pi$

You may be asked to find all the solutions in a given interval. To do this, find the principal value first. Then use the symmetries and periodicity of the graph to find further solutions in the the range.

You may be asked to give solutions in degrees or in radians. The following three examples are worked in degrees, with the answers in radians noted at the end of each part.

sin *x* = *c*

Example 4.17 Find the values of x in the interval $0° \leqslant x \leqslant 360°$ for which

a $\sin x = 0.5$

b $\sin x = -0.9$

Step 1: Find the principal value PV.

Step 2: Use a sketch of $y = \sin x$ to find other values in the given range.

a From calculator, PV $= \sin^{-1}(0.5) = 30°$.

The other value of x in the range is $180° - 30° = 150°$.

So $x = 30°, 150°$.

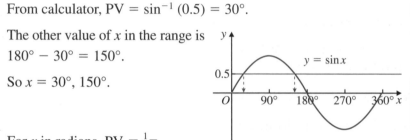

For x in radians, PV $= \frac{1}{6}\pi$.

In the interval $0 \leqslant x \leqslant 2\pi$, the solutions are $x = \frac{1}{6}\pi$ and $x = \pi - \frac{1}{6}\pi = \frac{5}{6}\pi$.

Tip:
Key in
[SHIFT] [SIN] [0.5] [=]

Tip:
Check on your calculator that $\sin x = 0.5$ for both values.

Tip:
Recognise this special angle (Section 4.5).

Calculator note:
If you are asked to give your answer as a multiple of π and you do not recognise it as a special angle (see Section 4.5), use your calculator in radian mode. Retaining the full display, divide by π and change the resulting decimal to a fraction by pressing the fraction key.

b From calculator, PV $= \sin^{-1}(-0.9) = -64.15...°$

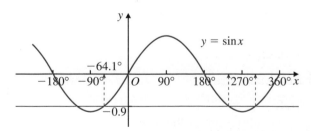

Other values of x *in range* are

$180° + 64.15...° = 244.15...°$

$360° - 64.15...° = 295.84...°$

So $x = 244°, 296°$ (nearest degree).

For x in radians, PV $= -1.12^c$. In the interval $0 \leqslant x \leqslant 2\pi$, the solutions are $x = \pi + 1.12^c$ and $x = 2\pi - 1.12^c$.

So $x = 4.26^c, 5.16^c$ (2 d.p.).

Recall:
For the sin function, the PV lies between $-90°$ and $90°$.

Tip:
Check on your calculator, but remember that you rounded, so you will not get $\sin x = -0.9$ exactly.

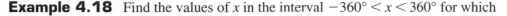

cos x = c

Example 4.18   Find the values of x in the interval $-360° < x < 360°$ for which

    a $\cos x = \dfrac{\sqrt{3}}{2}$ **b** $\cos x = -0.8$

Step 1: Find the principal value PV.

a From calculator, PV $= \cos^{-1}\left(\dfrac{\sqrt{3}}{2}\right) = 30°$.

Step 2: Use a sketch of $y = \cos x$ to find other values in the given range.

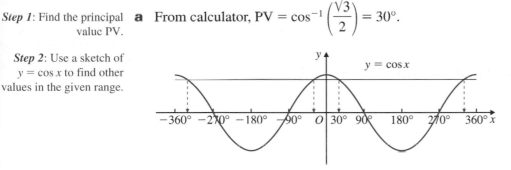

From the graph, other values of x are

$360° - 30° = 330°$, $-30°$ and
$-360° + 30° = -330°$.

So $x = -330°, -30°, 30°, 330°$.

The solutions can be written $x = +30°, \pm 330°$.

In radians: PV $= \frac{1}{6}\pi$.

Other solutions in the interval $-2\pi < x < 2\pi$ are $2\pi - \frac{1}{6}\pi$, $-\frac{1}{6}\pi$ and $-2\pi + \frac{1}{6}\pi$. So $x = \pm\frac{1}{6}\pi, \pm\frac{11}{6}\pi$.

Tip:
Check these on your calculator.

Tip:
Since the y-axis is a line of symmetry, solutions will always be of the form $\pm$.

b From calculator, PV $= \cos^{-1}(-0.8) = 143.13...°$.

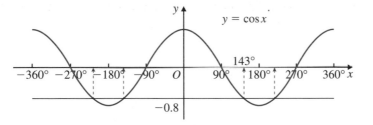

From the graph, other values of x are

$360° - 143.1...° = 216.8...°$,

$-143.13...°$ and $-216.8...°$.

So $x = \pm 143°, \pm 217°$ (nearest degree).

In radians: PV $= -2.5^{\text{c}}$ (1 d.p.).

Other solutions in the interval $-2\pi < x < 2\pi$ are $2\pi - 2.5^{\text{c}}$, -2.5^{c} and $-2\pi + 2.5^{\text{c}}$. So $x = \pm 2.5^{\text{c}}, \pm 3.8^{\text{c}}$.

Tip:
Use reflective symmetry in the y-axis.

tan x = c

Example 4.19   Find the values of x in the interval $-180° \leqslant x \leqslant 180°$ for which $\tan^2 x = 3$.

Step 1: Form equations in the form $\tan x = c$.

$\tan^2 x = 3 \Rightarrow \tan x = \sqrt{3}$ or $\tan x = -\sqrt{3}$

Consider $\tan x = \sqrt{3}$.

Tip:
There are two equations hidden in this question.

Step 2: For each equation, find the PV.

From calculator,
PV $= \tan^{-1}(\sqrt{3}) = 60°$.

Step 3: Use a sketch of $y = \tan x$ to find other values in the given range.

The other value of x in range is
$-180° + 60° = -120°$.

So $x = -120°, 60°$.

Now consider $\tan x = -\sqrt{3}$.

Note:
For the tan function, when you have found a solution in an interval of 180°, just add or subtract multiples of 180° to get further solutions.

From calculator,
PV $= \tan^{-1}(-\sqrt{3}) = -60°$.

The other value of x in range is
$180° - 60° = 120°$.

So $x = -60°, 120°$.

The complete solution is
$x = \pm 60°, \pm 120°$.

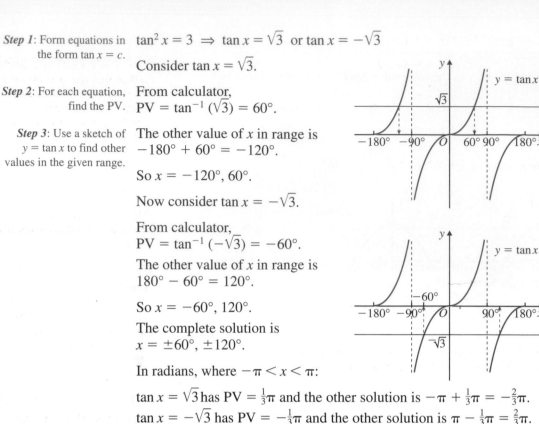

In radians, where $-\pi < x < \pi$:

$\tan x = \sqrt{3}$ has PV $= \frac{1}{3}\pi$ and the other solution is $-\pi + \frac{1}{3}\pi = -\frac{2}{3}\pi$.

$\tan x = -\sqrt{3}$ has PV $= -\frac{1}{3}\pi$ and the other solution is $\pi - \frac{1}{3}\pi = \frac{2}{3}\pi$.

So the complete solution is $x = \pm\frac{1}{3}\pi, \pm\frac{2}{3}\pi$.

Multiple angles

Take care if you are asked to solve equations involving multiples of x. This is illustrated in the following example.

Example 4.20 Find the values of x, in the interval $0° < x < 360°$, for which $\cos 2x = 0.5$.

Step 1: Make a substitution to get a simple equation.

Let $2x = \theta$, so the equation is $\cos \theta = 0.5$.

Step 2: Find the interval in which the new variable lies.

Interval required: $0° < x < 360°$

so $0° < 2x < 720° \Rightarrow 0° < \theta < 720°$

Step 3: Solve the equation in the new variable.

For θ, PV $= \cos^{-1}(0.5) = 60°$.

Other solutions in interval $0° < \theta < 720°$ are

$360° - 60°, 360° + 60°, 720° - 60°$.

So $\theta = 60°, 300°, 420°, 660°$.

Step 4: Substitute back for x.

$2x = 60°, 300°, 420°, 660°$

$\Rightarrow x = 30°, 150°, 210°, 330°$

Tip:
Do not divide by 2 until you have all the solutions for the new variable.

Using identities to solve trigonometric equations

Example 4.21 Find the values of θ, in the interval $-\pi < \theta < 2\pi$, for which $\sqrt{3}\sin\theta - \cos\theta = 0$, leaving your answers in terms of π.

Step 1: Rearrange to form an equation in $\tan\theta$, using an appropriate identity.

$$\sqrt{3}\sin\theta - \cos\theta = 0$$
$$\sqrt{3}\sin\theta = \cos\theta$$
($\div$ by $\cos\theta$) $\quad \sqrt{3}\dfrac{\sin\theta}{\cos\theta} = 1$
$$\sqrt{3}\tan\theta = 1$$
($\div$ by $\sqrt{3}$) $\quad \tan\theta = \dfrac{1}{\sqrt{3}}$

Note:
Before you attempt to divide by an unknown, make sure that it is not zero. In this case, $\cos\theta = 0$ is not a solution, so you can divide by it.

Step 2: Find the PV. $\quad$ PV $= \tan^{-1}\left(\dfrac{1}{\sqrt{3}}\right) = \frac{1}{6}\pi$

Step 3: Use the periodicity of the tan curve to find other solutions in the given interval.

Other solutions in range are

PV $+ \pi = \frac{1}{6}\pi + \pi = \frac{7}{6}\pi$

PV $- \pi = \frac{1}{6}\pi - \pi = -\frac{5}{6}\pi$

In the interval $-\pi < \theta < 2\pi$, $\quad \theta = -\frac{5}{6}\pi, \frac{1}{6}\pi, \frac{7}{6}\pi$.

Tip:
The tan function repeats every π radians, so add multiples of π to the PV.

Tip:
In radian mode, PV $= 0.5235...$. To write this as a multiple of π, divide by π and change the resulting decimal to a fraction.

Example 4.22 Find the values of θ in the interval $0 \le \theta \le 2\pi$ for which $2\sin^2\theta = \sin\theta$, leaving your answers in terms of π.

Step 1: Form an equation $f(\theta) = 0$ and factorise if possible.

$$2\sin^2\theta - \sin\theta = 0$$
$$\sin\theta(2\sin\theta - 1) = 0$$

Step 2: Solve the equations formed.

$\sin\theta = 0 \Rightarrow \theta = 0, \pi, 2\pi$

or $\quad 2\sin\theta - 1 = 0 \Rightarrow \sin\theta = \frac{1}{2}$

$$\theta = \frac{1}{6}\pi, \frac{5}{6}\pi$$

So $\theta = 0, \frac{1}{6}\pi, \frac{5}{6}\pi, \pi, 2\pi$.

Tip:
Do not cancel by $\sin\theta$, since $\sin\theta$ could be zero.

Example 4.23 **a** Write the expression $3 - 2\sin^2\theta - 3\cos\theta$ in terms of $\cos\theta$.

b Hence solve $3 - 2\sin^2\theta - 3\cos\theta = 0$ for values of θ in the interval $0 \le \theta \le 2\pi$, leaving your answers in terms of π where appropriate.

Step 1: Using an appropriate identity, form an equation in $\cos\theta$ only.

a $3 - 2\sin^2\theta - 3\cos\theta = 3 - 2(1 - \cos^2\theta) - 3\cos\theta$
$$= 3 - 2 + 2\cos^2\theta - 3\cos\theta$$
$$= 2\cos^2\theta - 3\cos\theta + 1$$

Tip:
Use $\sin^2\theta = 1 - \cos^2\theta$.

Step 2: Solve the equation in $\cos\theta$.

b $2\cos^2\theta - 3\cos\theta + 1 = 0$
$$(\cos\theta - 1)(2\cos\theta - 1) = 0$$
$\Rightarrow \quad \cos\theta - 1 = 0$
$$\cos\theta = 1$$
$$\theta = 0, 2\pi$$
or $\quad 2\cos\theta - 1 = 0$
$$\cos\theta = \frac{1}{2}$$
$$\theta = \frac{1}{3}\pi, 2\pi - \frac{1}{3}\pi$$
$$= \frac{1}{3}\pi, \frac{5}{3}\pi$$

Tip:
If the expression does not factorise, use the quadratic formula.

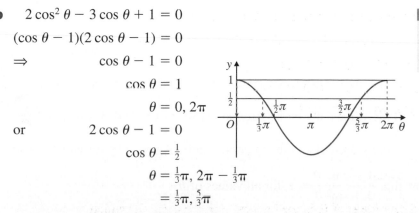

So, in the interval $0 \le \theta \le 2\pi$, $\theta = 0, \frac{1}{3}\pi, \frac{5}{3}\pi, 2\pi$.

Example 4.24 Find the values of x in the interval $0° < x < 180°$ for which $\sin(3x - 20°) = 0.2$. Give your answers correct to the nearest degree.

Step 1: Make a substitution to get a simple equation.

Let $3x - 20° = \theta$.
The equation becomes $\sin \theta = 0.2$.

Tip:
Work out the interval for $3x - 20°$.

Step 2: Find the interval in which the new variable lies.

Required interval: $0° < x < 180°$
$(\times 3)$ $0° < 3x < 540°$
$(-20°)$ $-20° < 3x - 20° < 520°$
Substitute θ $-20° < \theta < 520°$

Step 3: Work out the solutions for the new variable.

To solve $\sin \theta = 0.2$ for $-20° < \theta < 520°$, first find the PV.
$\text{PV} = \sin^{-1}(0.2) = 11.5...°$

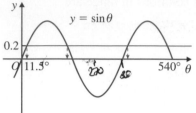

Other values of θ are

$180° - 11.5...° = 168.4...°$
$360° + 11.5...° = 371.5...°$
$540° - 11.5...° = 528.4...°$ (just out of range)

So $\theta = 11.5...°, 168.4...°, 371.5...°$.

Tip:
To maintain accuracy, work with uncorrected values if possible.

Tip:
Add 20°, then divide by 3 for each θ value.

Step 4: Write the solutions in terms of x.

$3x - 20° = 11.5...°, 168.4...°, 371.5...°$
$3x = 31.5...°, 188.4...°, 391.5...°$
$x = 10.5...°, 62.8...°, 130.5...°$

The values of x are $11°, 63°, 131°$ (correct to the nearest degree).

Tip:
Remember to give your answers to the required accuracy and in the given range.

SKILLS CHECK   **4C: Trigonometric functions and equations**

1 Solve the following equations for $0 \leqslant x < 360°$. If your answer is not exact, give it correct to the nearest degree.

    a $\sin x = 0.3$ **b** $\cos x = 0.5$ **c** $\tan x = -1.5$ **d** $\sin 2x = -0.5$

2 Solve the following equations for $0 \leqslant x \leqslant \pi$, leaving your answers in terms of π.

    a $\sin x = \dfrac{\sqrt{3}}{2}$ **b** $\cos x = -\dfrac{1}{\sqrt{2}}$ **c** $\tan x = -\sqrt{3}$ **d** $\cos 3x = 0.5$

3 On the same axes, sketch $y = \cos x$ and $y = \cos(x + 30°)$ where $0° \leqslant x \leqslant 360°$.

4 On the same axes, sketch $y = \sin x$ and $y = \sin(x - \frac{1}{4}\pi)$ for $0 \leqslant x \leqslant 2\pi$.

5 Find all the values of x in the interval $-180° \leqslant x \leqslant 180°$ for which $2\cos^2 x = \sqrt{3}\cos x$.

6 Find all the values of x in the interval $-360° \leqslant x \leqslant 360°$ for which $\sqrt{2}\sin^2 x - \sin x = 0$.

7 Show that $\dfrac{4 + \cos^2 \theta}{5 - \sin^2 \theta} \equiv 1$, for all values of θ.

8 Find the exact value of $\tan \theta$, given that $\sin \theta = -\frac{4}{5}$ and $\cos \theta = -\frac{3}{5}$.

9 a Given that $\sin 3x = \cos 3x$, write down the value of $\tan 3x$.

b Hence find all the solutions of the equation $\sin 3x = \cos 3x$ in the interval $0 < x < \pi$.

10 Find all the values of x in the interval $0 \leqslant x \leqslant 2\pi$ for which $2 \sin\left(x + \dfrac{\pi}{3}\right) = 1$, leaving your answers in terms of π.

11 a Given that $2 \sin^2 x = 1 - \cos x$, show that $2 \cos^2 x - \cos x - 1 = 0$.

b Hence find all the values of x in the interval $0 \leqslant x \leqslant 360°$ for which $2 \sin^2 x = 1 - \cos x$.

c **Write down** all the values of x in the interval $0 \leqslant x \leqslant 180°$ for which $2 \sin^2 2x = 1 - \cos 2x$.

SKILLS CHECK **4C EXTRA is on the CD**

Examination practice   Trigonometry

1 In triangle ABC, $AC = 50$ m, angle $BCA = 118°$ and angle $ABC = 35°$.

a Calculate the length of AB, giving your answer to the nearest metre.

b Calculate the area of triangle ABC.

2 To calculate the area of a field, $ABCD$, a farmer measures the boundary lengths and the length of a diagonal.

The measurements are

$AB = 350$ m $BC = 412$ m $CD = 729$ m
$DA = 295$ m $DB = 590$ m.

Calculate the area of the field, to the nearest hectare, where 1 hectare = $10\,000$ m^2.

3 In triangle PQR, angle $PQR = 150°$ and $PQ = 42$ cm. The area of the triangle is 630 cm^2.

Calculate

a length QR,

b length PR, giving your answer to the nearest mm.

4 Find all values of θ in the interval $0 \leqslant \theta \leqslant 360$ for which

a $\cos(\theta + 80)° = 0.5$,

b $\sin 2\theta° = 0.7$, giving your answers to one decimal place.

5 Triangle ABC has $AB = 9$ cm, $BC = 10$ cm and $CA = 5$ cm.

A circle, centre A and radius 3 cm, intersects AB and AC at P and Q respectively, as shown in the diagram.

a Show that, to 3 decimal places, $\angle BAC = 1.504$ radians.

Calculate

b the area, in cm^2, of the sector APQ,

c the area, in cm^2, of the shaded region $BPQC$,

d the perimeter, in cm, of the shaded region $BPQC$.

[Edexcel Jan 2001]

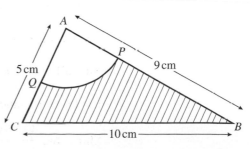

6 a Sketch, for $0 \leqslant x \leqslant 360°$, the graph of $y = \sin (x + 30°)$.

b Write down the coordinates of the points at which the graph meets the axes.

c Solve, for $0 \leqslant x \leqslant 360°$, the equation

$$\sin (x + 30°) = -\tfrac{1}{2}.$$

[Edexcel Jan 2003]

7 Find, in degrees, the value of θ in the interval $0 \leqslant \theta < 360°$ for which

$$2 \cos^2 \theta - \cos \theta - 1 = \sin^2 \theta.$$

Give your answers to 1 decimal place where appropriate.

[Edexcel June 2003]

8 $f(x) = 5 \sin 3x°, 0 \leqslant x \leqslant 180.$

a Sketch the graph of $f(x)$, indicating the value of x at each point where the graph intersects the x-axis.

b Write down the coordinates of all the maximum and minimum points of $f(x)$.

c Calculate the values of x for which $f(x) = 2.5$.

[Edexcel June 2002]

9 Find the values of θ, to 1 decimal place, in the interval $-180 \leqslant \theta \leqslant 180°$ for which

$$2 \sin^2 \theta° - 2 \sin \theta° = \cos^2 \theta°.$$

[Edexcel Jan 2002]

10 The curve C has equation $y = \cos \left(x + \dfrac{\pi}{4} \right), 0 \leqslant x \leqslant 2\pi$.

a Sketch C.

b Write down the exact coordinates of the points at which C meets the coordinate axes.

c Solve, for x in the interval $0 \leqslant x \leqslant 2\pi$,

$$\cos \left(x + \dfrac{\pi}{4} \right) = \dfrac{\sqrt{3}}{2},$$

giving your answers in terms of π.

11 i Solve, for $0° < x < 180°$, the equation

$$\sin (2x + 50°) = 0.6,$$

giving your answers to 1 decimal place.

ii In the triangle ABC, $AC = 18$ cm, $\angle ABC = 60°$ and $\sin A = \tfrac{1}{3}$.

a Use the sine rule to show that $BC = 4\sqrt{3}$.

b Find the exact value of $\cos A$.

[Edexcel Nov 2002]

5 Exponentials and logarithms

5.1 Exponential curves

y = aˣ and its graph.

The graph of $y = a^x$ is called an **exponential** curve. When $x = 0$, $y = a^0 = 1$, so the curve goes through $(0, 1)$.

The shape of the curve depends on the value of a.

When $a > 1$:

As $x \to \infty$, $y \to \infty$.

As $x \to -\infty$, $y \to 0$.

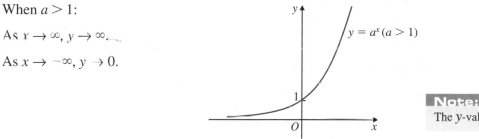

When $0 < a < 1$:

To get an idea of the general shape, let $a = \frac{1}{2}$ and consider $y = \left(\frac{1}{2}\right)^x$.

As $x \to \infty$, $y \to 0$.

As $x \to -\infty$, $y \to \infty$.

$\left(\frac{1}{2}\right)^x = 2^{-x}$, so when $a = \frac{1}{2}$, the curve is the graph of $y = 2^{-x}$. This is a reflection in the y-axis of $y = 2^x$.

> **Recall:**
> An exponent is another name for an index or a power (C1 Section 1.1).

> **Note:**
> The y-value is always positive.

> **Note:**
> Try substituting very large and very small numbers for x.

5.2 Logarithms

Logarithms and the laws of logarithms.

The **logarithm** (log) of a positive number to a given **base** is the **power** to which the base must be raised to equal the number.

$$y = a^x \iff x = \log_a y$$

The base a is such that $a > 0$ and $a \neq 1$.
For example:

$$5^3 = 125 \iff \log_5 125 = 3 \quad \text{(5 is the base)}$$

$$10^2 = 100 \iff \log_{10} 100 = 2 \quad \text{(10 is the base)}$$

> **Note:**
> $\iff$ indicates a two-way implication: $y = a^x \Rightarrow x = \log_a y$ and $x = \log_a y \Rightarrow y = a^x$.

Example 5.1 Find x in each of the following.

   a $\log_3 81 = x$ **b** $\log_3 1 = x$

   c $\log_3 3 = x$ **d** $\log_3 \left(\frac{1}{9}\right) = x$

a $\log_3 81 = x \Rightarrow 3^x = 81$
By inspection $x = 4$, since $3^4 = 81$.

b $\log_3 1 = x \Rightarrow 3^x = 1$
By inspection $x = 0$, since $3^0 = 1$.

c $\log_3 3 = x \Rightarrow 3^x = 3$
By inspection $x = 1$, since $3^1 = 3$.

d $\log_3 \left(\frac{1}{9}\right) = x \Rightarrow 3^x = \frac{1}{9}$
$$3^x = 3^{-2}$$
$$x = -2$$

> **Note:**
> For any base a, $\log_a 1 = 0$.

> **Recall:**
> $\frac{1}{9} = \frac{1}{3^2} = 3^{-2}$
> (see C1 Section 1.1).

Laws of logarithms

For any base a:
$$\log_a x + \log_a y = \log_a (xy)$$
$$\log_a x - \log_a y = \log_a \left(\frac{x}{y}\right)$$
$$k \log_a x = \log_a (x^k)$$

Examples:
$$\log_a 3 + \log_a 4 = \log_a 12$$
$$\log_a 20 - \log_a 5 = \log_a 4$$
$$2 \log_a 3 = \log_a 9$$

> **Tip:**
> Learn these laws and apply them accurately.

Special cases:
$$\log_a a = 1 \quad (\text{since } a^1 = a)$$
$$\log_a 1 = 0 \quad (\text{since } a^0 = 1)$$
$$\log_a (a^x) = x \log_a a = x$$
$$\log_a \left(\frac{1}{x}\right) = \log_a (x^{-1}) = -\log_a x$$

Examples:
$$\log_2 2 = 1$$
$$\log_3 1 = 0$$
$$\log_3 (3^4) = 4$$
$$\log_a \left(\tfrac{1}{3}\right) = -\log_a 3$$

Example 5.2 Find the value of x.

a $\log_a x = \log_a 20 - \log_a 15 + \log_a 3$

b $\log_a x = 2\log_a 8 - 3\log_a 4$

a $\log_a x = \log_a \left(\frac{20 \times 3}{15}\right) = \log_a 4$
$$\Rightarrow x = 4$$

b $\log_a x = 2\log_a 8 - 3\log_a 4$
$$= \log_a 8^2 - \log_a 4^3$$
$$= \log_a 64 - \log_a 64$$
$$= 0$$
$$\Rightarrow x = a^0 = 1$$

Example 5.3 Simplify $\log_a (a\sqrt{a})$.

$\log_a(a\sqrt{a}) = \log_a a + \log_a (a^{\frac{1}{2}})$
$$= 1 + \tfrac{1}{2} \log_a a$$
$$= 1 + \tfrac{1}{2}$$
$$= 1\tfrac{1}{2}$$

> **Tip:**
> You could write $a\sqrt{a} = a^{\frac{3}{2}}$; then $\log_a(a^{\frac{3}{2}}) = \frac{3}{2} \log_a a = \frac{3}{2}$.

Example 5.4 **a** Given $\log_2 16 = x$, find x.

b Write down the value of   **i** $\log_2 16^3$ **ii** $\log_2 \dfrac{1}{16^2}$.

Step 1: Evaluate the log. **a** $\log_2 16 = x \Rightarrow 2^x = 16 \Rightarrow x = 4$

Step 2: Simplify using the log laws.

b i $\log_2 16^3 = 3(\log_2 16) = 3 \times 4 = 12$

ii $\log_2 \dfrac{1}{16^2} = \log_2 16^{-2} = -2(\log_2 16) = -2 \times 4 = -8$

Example 5.5 Given that $\log_3 x = \log_9 3$, find the value of x.

Step 1: Evaluate the right-hand side.

Consider the right-hand side:

Let $\log_9 3 = y$

Then $9^y = 3$

$(3^2)^y = 3^1$

Recall:
$(a^m)^n = a^{mn}$ (C1 Section 1.1).

Equating indices:

$2y = 1$

$y = \tfrac{1}{2}$

Step 2: Convert to exponential form.

So $\log_3 x = \tfrac{1}{2}$

$\Rightarrow \qquad x = 3^{\frac{1}{2}} = \sqrt{3}$

5.3 Exponential equations

The solution of equations of the form $a^x = b$.

One way of solving equations of the form $a^x = b$ is to take logs to the base 10 of both sides.

Example 5.6 Solve **a** $5^x = 51$ **b** $2^{3x+1} = 12$.

Step 1: Take logs to base 10 of both sides.

Step 2: Simplify using the log laws.

Step 3: Solve the equation in x.

a $\log_{10}(5^x) = \log_{10} 51$

$x \log_{10} 5 = \log_{10} 51$

$x = \dfrac{\log_{10} 51}{\log_{10} 5}$

$= 2.44$ (3 s.f.)

Note:
Use $\boxed{\log}$ on your calculator. This is programmed to give logs to base 10.

b $\log_{10}(2^{3x+1}) = \log_{10} 12$

$(3x + 1)\log_{10} 2 = \log_{10} 12$

$3x + 1 = \dfrac{\log_{10} 12}{\log_{10} 2}$

$= 3.584...$

$3x = 3.584... - 1 = 2.584...$

$x = 0.862$ (3 s.f.)

Tip:
These logs are divided, not subtracted so do not try to cancel here.

Change-of-base formula

You can write a log in terms of a different base by using the following formula:

$$\log_a b = \frac{\log_c b}{\log_c a}, \quad \text{for any constant } c$$

It then possible to evaluate a logarithm on the calculator by choosing $c = 10$. For example:

Tip:
Use $\boxed{\log}$ on your calculator.

$$\log_3 2 = \frac{\log_{10} 2}{\log_{10} 3} \quad \text{(substituting } c = 10)$$

$$= 0.631 \text{ (3 s.f.)}$$

Note:
Writing the log in index form gives $3^{0.631\ldots} = 2$.

Example 5.7 Given that $\log_2 x = 9 \log_x 2$, find x.

Step 1: Use the change-of-base formula to ensure the bases of the logs are the same in the equation.

Write $\log_x 2$ in terms of a log to the base 2, so that the bases are then the same. Substituting $c = 2$ into the change-of-base formula, you get

$$\log_x 2 = \frac{\log_2 2}{\log_2 x} = \frac{1}{\log_2 x} \quad \text{(since } \log_2 2 = 1)$$

Note:
You can choose either term to change the base.

The equation now becomes

$$\log_2 x = 9 \frac{1}{\log_2 x}$$

Step 2: Rewrite the common log term as m and solve for m.

Let $m = \log_2 x$

$$m = \frac{9}{m}$$
$$m^2 = 9$$
$$m = \pm 3$$

Step 3: Substitute back for m to find x.

So, $\log_2 x = 3$ $\qquad$ or $\quad \log_2 x = -3$

$\qquad\qquad x = 2^3 = 8$ $\qquad\qquad\qquad x = 2^{-3} = \frac{1}{8}$

Hence $x = 8$ or $\frac{1}{8}$.

Example 5.8 Solve the following pair of simultaneous equations where $x > 0$, $y > 0$.

$$\log_{10} x = 1 - \log_{10} y$$
$$\log_2(2x - y) = 3$$

Note:
lg is sometimes used as shorthand for $\log_{10}$.

Step 1: Bring the log terms to one side of the equation and combine using the log rules if necessary.

$$\log_{10} x = 1 - \log_{10} y$$
$$\log_{10} x + \log_{10} y = 1$$
$$\log_{10} (xy) = 1$$

$$\log_2(2x - y) = 3$$

Tip:
Check that the bases are the same for the logs in each equation.

Step 2: Use $x = \log_a y \Leftrightarrow y = a^x$ to eliminate the logs.

$$xy = 10^1 \qquad\qquad \textcircled{1}$$
$$2x - y = 2^3 = 8 \qquad \textcircled{2}$$

Step 3: Solve for the unknowns.

From $\textcircled{2}$: $\qquad\qquad\qquad\qquad y = 2x - 8$

Substitute into $\textcircled{1}$: $\qquad x(2x - 8) = 10$
$$2x^2 - 8x - 10 = 0$$
$$x^2 - 4x - 5 = 0$$
$$(x - 5)(x + 1) = 0$$
$$x = 5, x = -1$$

Recall:
Factorising quadratics in C1 Section 1.5.

Since $x > 0$, $x = 5$ and hence $y = 2x - 8 = 2$.

The solution to the equation is $x = 5$ and $y = 2$.

1 Evaluate **a** $\log_4 64$ **b** $\log_5 25$ **c** $\log_2 8$ **d** $\log_{16} 4$.

2 Evaluate **a** $\log_2 8^3$ **b** $\log_3 (\tfrac{1}{9})$ **c** $\log_4 \sqrt{64}$ **d** $\dfrac{\log_3 27}{\log_3 9}$.

3 Find the value of x.

    a $\log_a x = \log_a 30 - \log_a 5 - \log_a 3$ **b** $\log_a x = 2\log_a 2 + 2\log_a 3$

4 Simplify **a** $\log_a a^5$ **b** $\log_a \left(\dfrac{1}{\sqrt{a}}\right)$ **c** $4\log_a 1 + 3\log_a a$.

5 Express $\log_2 \sqrt{\dfrac{p^2 q}{2r^3}}$ in terms of $\log_2 p$, $\log_2 q$ and $\log_2 r$.

6 Given that $\log_a x = 2(\log_a 3 + \log_a 2)$, where a is a positive constant, find x.

7 **a** Write down the value of **i** $\log_3 3$ **ii** $\log_3 27$.

    b Find the value of $\log_3 2 - \log_3 54$.

8 Solve the following, giving your answers correct to three significant figures.

    a $2^x = 27$ **b** $3^{5x-2} = 20$ **c** $2\log_x 5 = 3$ **d** $x = \log_5 10$ **e** $x = \log_3 7$

9 **a** Show that $\log_a b + \log_a b^2 + \log_a b^3 + \cdots$ $(b \neq 1)$ is an arithmetic series and state the common difference of the series. (Hint: see C1 Section 3.3 for arithmetic series.)

    b The sum of the first ten terms of the series is $k \log_a b$. Find k.

10 It is given that $3^{2x} = 10(3^x) - 9$.

    a Writing $y = 3^x$, show that $y^2 - 10y + 9 = 0$.

    b Solve $3^{2x} = 10(3^x) - 9$.

11 Given that $\log_3 x + 6 \log_x 3 = 5$, find x.

12 Solve the following pair of simultaneous equations where $p > 0$, $q > 0$.

$$\log_{10} p = 4 - \log_{10} q$$
$$\log_2 (p - 2q) = 1$$

SKILLS CHECK **5A EXTRA is on the CD**

Examination practice Exponentials and logarithms

1 Every £1 of money invested in a savings scheme continuously gains interest at a rate of 4% per year. Hence, after x years, the total value of an initial £1 investment is £y, where

$$y = 1.04^x.$$

    a Sketch the graph of $y = 1.04^x$, $x \geq 0$.

    b Calculate, to the nearest £, the total value of an initial £800 investment after 10 years.

    c Use logarithms to find the number of years it takes to double the total value of any initial investment.

                                                 [Edexcel Nov 2003]

2 Given that $p = \log_q 16$, express in terms of p.

   a $\log_q 2$.

   b $\log_q (8q)$.

[Edexcel Jan 2002]

3 **a** Simplify $\dfrac{x^2 + 4x + 3}{x^2 + x}$.

   b Find the value of x for which $\log_2 (x^2 + 4x + 3) - \log_2 (x^2 + x) = 4$.

[Edexcel June 2003]

4 Solve, giving your answers as exact fractions, the simultaneous equations

$$8^y = 4^{2x + 3},$$

$$\log_2 y = \log_2 x + 4.$$

[Edexcel June 2000]

5 **a** Using the substitution $u = 2^x$, show that the equation $4^x - 2^{(x + 1)} - 15 = 0$ can be written in the form $u^2 - 2u - 15 = 0$

   b Hence solve the equations $4^x - 2^{(x + 1)} - 15 = 0$, giving your answers to 2 decimals places.

[Edexcel Nov 2002]

6 The sequence $u_1, u_2, u_3, \ldots, u_n$ is defined by the recurrence relation

$$u_{n + 1} = pu_n + 5, u_1 = 2, \text{ where } p \text{ is a constant.}$$

Given that $u_3 = 8$,

   a show that one possible value of p is $\frac{1}{2}$ and find the other value of p.

Using $p = \frac{1}{2}$,

   b write down the value of $\log_2 p$.

Given also that $\log_2 q = t$,

   c express $\log_2 \left(\dfrac{p^3}{\sqrt{q}} \right)$ in terms of t.

[Edexcel Nov 2002]

7 **a** Given that $3 + 2\log_2 x = \log_2 y$, show that $y = 8x^2$.

   b Hence, or otherwise, find the roots α and β, where $\alpha < \beta$, of the equation

$$3 + 2\log_2 x = \log_2 (14x - 3).$$

   c Show that $\log_2 \alpha = -2$.

   d Calculate $\log_2 \beta$, giving your answer to 3 significant figures.

[Edexcel June 2002]

8 Find the value of x if $3^{x - 2} = 2^{x + 1}$. Give your answer correct to one decimal place.

9 **a** Given that $2 + \log_3 x = \log_3 y$, show that $y = 9x$.

   b Hence, or otherwise, solve $2 + \log_3 x = \log_3 (5x + 2)$.

10 **a** Simplify $\dfrac{x^2 + 5x + 6}{x^2 + 3x + 2}$.

   b Find the value of x for which $\log_3 (x^2 + 5x + 6) = 2 + \log_3 (x^2 + 3x + 2)$.

6 Differentiation

In module C1 you differentiated terms of the form ax^n.

The rule is as follows:

For any rational number n, $y = ax^n \Rightarrow \dfrac{dy}{dx} = anx^{n-1}$

Note:
n can be positive or negative, an integer or a fraction.

Recall:
Multiply by the power of x and decrease the power by 1.

You also need to remember the following:

$$y = f(x) \pm g(x) \Rightarrow \frac{dy}{dx} = f'(x) \pm g'(x)$$

Recall:
This means that you can differentiate term by term.

The second derivative, $\dfrac{d^2y}{dx^2}$, is obtained by differentiating $\dfrac{dy}{dx}$ with respect to x. In function notation, if $y = f(x)$, the second derivative is written $f''(x)$.

6.1 Stationary points

Applications of differentiation to maxima, minima and stationary points.
Second order derivatives.

At a **stationary point** on a curve, the gradient is zero, so $\dfrac{dy}{dx} = 0$.

There are three types of stationary points: maximum turning points; minimum turning points; and stationary points of inflexion.

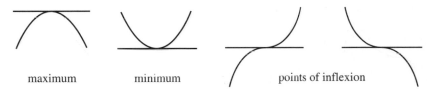

maximum   minimum   points of inflexion

Note:
The tangent at a stationary point is parallel to the x-axis.

Determining the nature of a stationary point

Here are two methods for determining the nature of a stationary point:

Note:
Another method involves investigating the y-value immediately to the left and right of the stationary point.

Method 1

Investigate the value of $\dfrac{dy}{dx}$ for x-values immediately to the left and to the right of the point.

If, as x increases,

- $\dfrac{dy}{dx}$ goes from positive to zero to negative, there is a maximum point

- $\dfrac{dy}{dx}$ goes from negative to zero to positive, there is a minimum point

- $\dfrac{dy}{dx}$ goes from positive to zero to positive, or negative to zero to negative, there is a stationary point of inflexion.

Note:
Direction of slope near stationary point:

Maximum:

Minimum:

Point of inflexion:

or

Method 2

Consider the sign of $\dfrac{d^2y}{dx^2}$ at the stationary point:

- $\dfrac{dy}{dx} = 0$ and $\dfrac{d^2y}{dx^2} > 0 \Rightarrow$ minimum turning point

- $\dfrac{dy}{dx} = 0$ and $\dfrac{d^2y}{dx^2} < 0 \Rightarrow$ maximum turning point.

Note that if $\dfrac{dy}{dx} = 0$ and $\dfrac{d^2y}{dx^2} = 0$, Method 2 fails and Method 1 is advised.

Note:

$\dfrac{d^2y}{dx^2} = 0$ does not imply that the stationary point is a point of inflexion.

Example 6.1

a Find the coordinates of the stationary points on the curve $y = x^3 - 3x + 2$.

b Investigate their nature. **c** Sketch the curve.

Tip:
Notice that the question suggests there is more than one.

Step 1: Find $\dfrac{dy}{dx}$.

a $y = x^3 - 3x + 2 \Rightarrow \dfrac{dy}{dx} = 3x^2 - 3$

Step 2: Put $\dfrac{dy}{dx} = 0$ and solve for x.

$\dfrac{dy}{dx} = 0$ when $3x^2 - 3 = 0$

$3(x - 1)(x + 1) = 0$

$\Rightarrow \quad x = 1$ or $x = -1$

Step 3: Substitute x-value into equation of the curve.

When $x = 1$, $y = 1^3 - 3(1) + 2 = 0$.
When $x = -1$, $y = (-1)^3 - 3(-1) + 2 = 4$.

There are stationary points at $(1, 0)$ and $(-1, 4)$.

Tip:
Take care with negatives.

Step 4: Find $\dfrac{d^2y}{dx^2}$ and substitute x-value to find nature.

b Using the second derivative:

$\dfrac{dy}{dx} = 3x^2 - 3 \Rightarrow \dfrac{d^2y}{dx^2} = 6x$

When $x = 1$, $\dfrac{d^2y}{dx^2} = 6(1) = 6 > 0 \Rightarrow$ minimum point.

When $x = -1$, $\dfrac{d^2y}{dx^2} = 6(-1) = -6 < 0 \Rightarrow$ maximum point.

There is a minimum turning point at $(1, 0)$ and a maximum turning point at $(-1, 4)$.

Tip:
Summarise the information, so that it is easier for the examiner to see.

c To sketch the curve, you also need axes intercepts. You can check using the Factor Theorem that the equation of the curve y can be written as $y = (x - 1)^2(x + 2)$.

Recall:
The Factor Theorem, Section 1.1.

Step 5: Set $x = 0$ to find the y-intercepts and $y = 0$ to find the x-intercepts.

When $x = 0$, $y = 2$; when $y = 0$, $x = 1$ or $x = -2$. So the curve crosses the y-axis at $(0, 2)$ and the x-axis at $(1, 0)$ and $(-2, 0)$.

Recall:
Shapes of cubic curves (C1 Section 1.11).

Step 6: Use your information to sketch the curve.

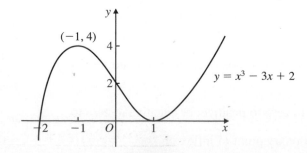

Example 6.2 Find the coordinates of the stationary point on the curve $y = x^3 - 3x^2 + 3x - 1$ and state its nature. Sketch the curve, showing the intercepts with the coordinate axes.

Step 1: Find $\dfrac{dy}{dx}$.

$y = x^3 - 3x^2 + 3x - 1 \Rightarrow \dfrac{dy}{dx} = 3x^2 - 6x + 3$

Step 2: Set $\dfrac{dy}{dx} = 0$ and solve for x.

$\dfrac{dy}{dx} = 0$ when $3x^2 - 6x + 3 = 0$

$(\div 3)$ $x^2 - 2x + 1 = 0$

$(x - 1)(x - 1) = 0$

$\Rightarrow$ $x = 1$

Step 3: Substitute the x-value into the equation of the curve.

When $x = 1$, $y = 1^3 - 3(1^2) + 3(1) - 1 = 0$.
There is a stationary point at $(1, 0)$.

Step 4: Find $\dfrac{d^2y}{dx^2}$ and substitute the x-value to find nature.

Using the second derivative:

$\dfrac{dy}{dx} = 3x^2 - 6x + 3 \Rightarrow \dfrac{d^2y}{dx^2} = 6x - 6$

When $x = 1$, $\dfrac{d^2y}{dx^2} = 6(1) - 6 = 0$.

Since $\dfrac{d^2y}{dx^2} = 0$, the second derivative method has broken down, so use Method 2 near $x = 1$.

When $x = 0$, $\dfrac{dy}{dx} = 3(0^2) - 6(0) + 3 = 3$.

Step 5: Calculate the value of $\dfrac{dy}{dx}$ close to the stationary point.

When $x = 2$, $\dfrac{dy}{dx} = 3(2^2) - 6(2) + 3 = 3$.

Step 6: Represent the gradients in a table.

	$x = 0$	$x = 1$	$x = 2$
Sign of $\dfrac{dy}{dx}$	$+$	0	$+$
Direction of gradient	/	—	/

There is a stationary point of inflexion at $(1, 0)$.

Step 7: State the nature of the stationary point.

When $x = 0$, $y = 0^3 - 3(0)^2 + 3(0) - 1 = -1$.
So the curve crosses the y-axis at $(0, -1)$.

Step 8: Set $x = 0$ to find the y-intercept.

Sketch of $y = x^3 - 3x^2 + 3x - 1$:

Step 9: Use your information to sketch the curve.

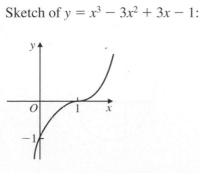

Example 6.3 A curve has equation $y = \dfrac{1}{x} + 32x^2$. Find the coordinates of the stationary point on the curve and determine its nature.

Step 1: Write terms in the form ax^n using the index laws.

$$y = \frac{1}{x} + 32x^2 = x^{-1} + 32x^2$$

Tip:
Write in index form before differentiating.

Step 2: Find $\dfrac{\mathrm{d}y}{\mathrm{d}x}$.

$$\frac{\mathrm{d}y}{\mathrm{d}x} = -x^{-2} + 64x = -\frac{1}{x^2} + 64x$$

Step 3: Set $\dfrac{\mathrm{d}y}{\mathrm{d}x} = 0$ and solve for x.

$$\frac{\mathrm{d}y}{\mathrm{d}x} = 0 \text{ when } -\frac{1}{x^2} + 64x = 0$$

$$64x = \frac{1}{x^2}$$

$$x^3 = \tfrac{1}{64}$$

$$x = \sqrt[3]{\tfrac{1}{64}} = \tfrac{1}{4}$$

Tip:
Multiply both sides by x^2 and divide by 64.

Step 4: Substitute the x-value into the equation of the curve.

When $x = \tfrac{1}{4}$, $y = \dfrac{1}{x} + 32x^2 = \dfrac{1}{\frac{1}{4}} + 32 \times (\tfrac{1}{4})^2 = 6$.

Hence there is a stationary point at $(\tfrac{1}{4}, 6)$.

Tip:
You must include sufficient working to show whether the second differential is positive or negative.

Step 5: Find $\dfrac{\mathrm{d}^2y}{\mathrm{d}x^2}$ and substitute the x-value to find nature.

$$\frac{\mathrm{d}y}{\mathrm{d}x} = -x^{-2} + 64x \Rightarrow \frac{\mathrm{d}^2y}{\mathrm{d}x^2} = 2x^{-3} + 64$$

When $x = \tfrac{1}{4}$, $\dfrac{\mathrm{d}^2y}{\mathrm{d}x^2} = 2 \times (\tfrac{1}{4})^{-3} + 64 = 192$.

Since $\dfrac{\mathrm{d}^2y}{\mathrm{d}x^2} > 0$, $(\tfrac{1}{4}, 6)$ is a minimum turning point.

6.2 Increasing and decreasing functions

Applications of differentiation to increasing and decreasing functions.

Consider a function f(x).

If, as the x-value increases, the corresponding value of f(x) increases, the function is an **increasing function**.

The gradient of the curve $y = $ f(x) is positive.

If, as the x-value increases, the corresponding value of f(x) decreases, the function is a **decreasing function**.

The gradient of the curve $y = $ f(x) is negative.

Note:
Curves can be both increasing and decreasing, depending on which x-values you are considering.

Example 6.4 Find the set of values for which the function f(x) = $x^3 - x^2 - x + 4$ is decreasing.

Step 1: Find f$'(x)$.

f(x) = $x^3 - x^2 - x + 4 \Rightarrow$ f$'(x) = 3x^2 - 2x - 1$

Step 2: Set f$'(x) < 0$.

For a decreasing function, f$'(x) < 0$
$$\Rightarrow 3x^2 - 2x - 1 < 0$$

Step 3: Solve the inequality.

$$(3x + 1)(x - 1) < 0$$
$$\Rightarrow \tfrac{1}{3} < x < 1$$

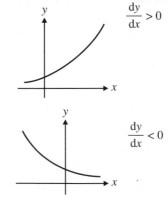

Recall:
Solving quadratic inequalities (C1 Section 1.9).

6.3 Application problems

Application to determining maxima and minima.

Differentiation can be used to solve practical problems involving maximum and minimum values. In these, you consider two variables representing some measures that are related, such as the length of a rectangle and its area. Allow one to vary and use differentiation to find the maximum or minimum value of the other.

Note:
These are sometimes called optimisation problems.

Example 6.5 A carpenter is building a rectangular tabletop. In order to ensure the correct sizing he wishes to have a perimeter of 200 cm. Given that he can vary the length, he wants to know the maximum area the tabletop can have.

a By letting x cm be the length of one side of the tabletop, form an expression in terms of x, for A, the area in cm^2, of the tabletop.

b Use differentiation to find the maximum area of the top, verifying that you have found the maximum.

c Describe the rectangular tabletop that has maximum area.

Step 1: If applicable, draw a clear diagram, labelling all the measures and identify any unknown measures.

a

x cm

h cm
(unknown)

Tip:
Let the unknown measure be h.

Tip:
Use the information given about the object to express h in terms of x.

Step 2: Express any unknown measures in terms of x.

Perimeter $= 200$ cm
$\Rightarrow 2x + 2h = 200$
$\qquad 2h = 200 - 2x$
$\qquad\; h = 100 - x \qquad$ ①

Step 3: Form an expression for A in terms of x.

b Let the area be A cm^2.
$A = x \times h$
Substituting for h from ①:
$A = x(100 - x)$
$\quad = 100x - x^2$

Note:
This expression tells you how A changes for different values of x.

Step 4: Set $\dfrac{dA}{dx} = 0$ and solve for x.

$\dfrac{dA}{dx} = 100 - 2x$

$\dfrac{dA}{dx} = 0$ when $100 - 2x = 0 \Rightarrow x = 50$

Note:
This is the value of the length that gives a stationary point for A.

Step 5: Find $\dfrac{d^2A}{dx^2}$ and substitute the x-value to find nature.

$\dfrac{d^2A}{dx^2} = -2$

When $x = 50$, $\dfrac{d^2A}{dx^2} = -2 < 0 \Rightarrow$ maximum value of A.

Note:
In this case, $\dfrac{d^2A}{dx^2} < 0$ for all values of x.

Step 6: Substitute the x-value into the equation for A.

$A = 100 \times 50 - 50^2 = 2500$

The maximum area of the tabletop is 2500 cm^2 and this occurs when the length is 50 cm.

Step 7: Consider the dimensions of the tabletop.

c Length $= x$ cm $= 50$ cm
Width $= (100 - x)$ cm $= 50$ cm

The rectangular tabletop that gives the maximum area is in the shape of a square.

Example 6.6   A cylindrical tin, closed at both ends, is made from thin sheet metal. The radius of the base of the cylinder is r cm and the volume of the tin is 1024π cm^3.

a Show that the total surface area, S cm^2, of the cylinder is given by
$$S = \frac{2048\pi}{r} + 2\pi r^2.$$

b Find the value of r that gives a minimum total surface area and state the value of this surface area, in terms of π.

Step 1: Draw a diagram showing given information.

Step 2: Write unknown measures in terms of the given variable.

a   Let the height be h cm.

Volume $= 1024\pi$

$\Rightarrow \pi r^2 h = 1024\pi$

$\quad h = \dfrac{1024}{r^2}$

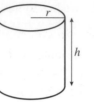

Tip:
Use the condition that the volume is 1024π to express h in terms of r.

Step 3: Find an expression for S in terms of R.

$S = 2\pi r h + 2\pi r^2$

$\quad = 2\pi r \left(\dfrac{1024}{r^2}\right) + 2\pi r^2$

$\quad = \dfrac{2048\pi}{r} + 2\pi r^2$

Tip:
Find the curved surface area and the area of the two circular ends.

Step 4: Set $\dfrac{\mathrm{d}S}{\mathrm{d}r} = 0$ and solve for r.

b $S = 2048\pi r^{-1} + 2\pi r^2$

$\dfrac{\mathrm{d}S}{\mathrm{d}r} = -2048\pi r^{-2} + 4\pi r$

$\dfrac{\mathrm{d}S}{\mathrm{d}r} = 0$ when $-2048\pi r^{-2} + 4\pi r = 0$

$$4\pi r = \frac{2048\pi}{r^2}$$

$(\times r^2)$ $\qquad\qquad\qquad 4\pi r^3 = 2048\pi$

$(\div 4\pi)$ $\qquad\qquad\qquad r^3 = \dfrac{2048\pi}{4\pi} = 512$

$$r = \sqrt[3]{512} = 8$$

There is a stationary value when $r = 8$.

Tip:
Write all terms in index form before differentiating.

Step 5: Find $\dfrac{\mathrm{d}^2 S}{\mathrm{d}r^2}$ and substitute the x-value to find nature.

$\dfrac{\mathrm{d}S}{\mathrm{d}r} = -2048\pi r^{-2} + 4\pi r \Rightarrow \dfrac{\mathrm{d}^2 S}{\mathrm{d}r^2} = 4096\pi r^{-3} + 4\pi$

When $r = 8$, $\dfrac{\mathrm{d}^2 S}{\mathrm{d}r^2} = 4096\pi \times 8^{-3} + 4\pi = 37.6\ldots > 0$

So S has a minimum value when $r = 8$.

Step 6: Substitute for r into S.

When $r = 8$, $S = \dfrac{2048\pi}{8} + 2\pi \times 8^2 = 384\pi$.

The minimum surface area is 384π cm^2.

Tip:
Remember to give your answer in terms of π and include the units.

1 Find the coordinates of the stationary points of the given curves and determine their nature. Sketch the curves for parts **a** and **b**.

  a $y = x^2 + x^3$ **b** $y = x^3 - 3x$

  c $y - 2x^3 + 3x^2 - 12x + 6$ **d** $y - 3x^4 - 8x^3 + 6x^2 - 3$

2 Find the x-coordinates of the stationary points on the curve $y = (1 - x^2)(1 - 4x)$ and determine their nature.

3 The diagram shows a sketch of the curve $y = 12x^{\frac{1}{2}} - 2x^{\frac{3}{2}}$.

  a Find the x-coordinate of B.

  b Hence state the maximum value of y, leaving your answer in the form $a\sqrt{2}$, where a is an integer to be found.

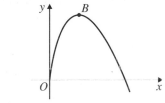

4 Find the set of values of x for which the function is **i** increasing **ii** decreasing:

  a $f(x) - (x - 2)^2$ **b** $f(x) = x^2 + 8$ **c** $g(x) = x^3 - 3x + 2$

5 It is given that $f(x) = (2x - 1)(3x + 1)$. Find the values of x for which $f(x)$ is an increasing function.

6 A builder wishes to make a rectangular enclosure around a garden. The house is to form one of the sides; this side has length $4x$ metres. The other three sides are to be fenced with a total of 2000 m of fencing.

  Show that the area of the garden, A, is given by $A = 4000x - 8x^2$.

  Hence, find the maximum area of the garden, verifying that the value you have found is a maximum.

7 The sum of two variable positive numbers is 200.

  Let x be one of the numbers, and let the product of these two numbers be y.

  Find the maximum value of y.

8 A closed cuboid is to be made from thin cardboard. The base of the cuboid is a rectangle with width x cm. The length of the base is twice the width and the volume of the cuboid is 1944 cm^3. The surface area of the cuboid is S cm^2.

  a Show that $S = 4x^2 + 5832x^{-1}$.

  b Given that x can vary, find the value of x that makes the surface area a minimum.

  c Find the minimum value of the surface area.

9 The diagram shows a prism where the cross-section is in the form of a sector OPQ of a circle, centre O and radius x cm. The length of the prism is $2x$ cm and the angle POQ is θ radians. The volume of the prism is 4608 cm^3.

  a Show that $\theta = \dfrac{4608}{x^3}$.

  b Show that the surface area, S cm^2, of the prism is given by $S = 4x^2 + \dfrac{13\,824}{x}$.

  c Given that x can vary, find the minimum value of the surface area of the prism, verifying that you have found the minimum value.

1 The function f, defined for $x \in \mathbb{R}$, $x > 0$, is such that

$$f(x) = \frac{1}{3}x^3 + 2x - \frac{1}{x} - \frac{8}{3}$$

  a Find the value of $f''(x)$ at $x = 4$.

  b Prove that f is an increasing function.

2 On a journey, the average speed of a car is v m s^{-1}. For $v \geqslant 5$, the cost per kilometre, C pence, of the journey is modelled by

$$C = \frac{160}{v} + \frac{v^2}{100}.$$

Using this model,

  a show, by calculus, that there is a value of v for which C has a stationary value, and find this value of v.

  b Justify that this value of v gives a minimum value of C.

  c Find the minimum value of C and hence find the minimum cost of a 250 km car journey.

[Edexcel Jan 2003]

3 A container made from thin metal is in the shape of a right circular cylinder with height h cm and base radius r cm. The container has no lid. When full of water, the container holds 500 cm^3 of water.

$$A = \pi r^2 + \frac{1000}{r}.$$

  a Show that the exterior surface area, A cm^2, of the container is given by

  b Find the value of r for which A is a minimum.

  c Prove that this value of r gives a minimum value of A.

  d Calculate the minimum value of A, giving your answer to the nearest integer.

[Edexcel Nov 2003]

4 An architect is drawing up plans for a mini-theatre. The diagram shows the plan of the base which consists of a rectangle of length $2y$ metres and width $2x$ metres and a semicircle of radius x metres which is placed with one side of the rectangle as diameter.

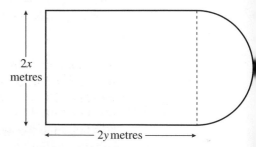

Find, in terms of x and y, expressions for

  a the perimeter of the base,

  b the area of the base.

The architect decides the base should have a perimeter of 100 metres.

  c Show that the area A square metres of the base is given by

$$A = 100x - 2x^2 - \tfrac{1}{2}\pi x^2.$$

  d Given that x can vary, find the value of x for which $\frac{dA}{dx} = 0$ and determine the corresponding value of y, giving your answers to 2 significant figures.

  e Find the maximum value of A and explain why this value is a maximum.

[Edexcel Specimen Paper]

5 The curve with equation $y = (2x + 1)(x^2 - k)$, where k is a constant, has a stationary point where $x = 1$.

  a Determine the value of k.

  b Find the coordinates of the stationary points and determine the nature of each.

  c Sketch the curve and mark on your sketch the coordinates of the points where the curve crosses the coordinate axes.   [Edexcel Mock Paper]

6 For the curve C with equation $y = x^4 - 8x^2 + 3$,

  a find $\dfrac{dy}{dx}$.

  b find the coordinates of each of the stationary points,

  c determine the nature of each stationary point.

The point A, on the curve C, has x-coordinate 1.

  d Find an equation for the normal to C at A, giving your answer in the form $ax + by + c = 0$, where a, b and c are integers.   [Edexcel June 2003]

7 A large tank in the shape of a cuboid is to be made from 54 m^2 of sheet metal. The tank has a horizontal rectangular base and no top. The height of the tank is x metres. Two of the opposite vertical faces are squares.

  a Show that the volume, $V \text{ m}^3$, of the tank is given by

$$V = 18x - \frac{2}{3}x^3.$$

  b Given that x can vary, use differentiation to find the maximum value of V.

  c Justify that the value of V you have found is a maximum.   [Edexcel May 1995]

8

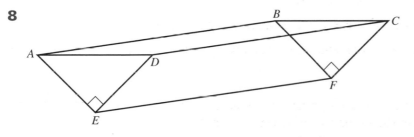

The diagram shows an open tank for storing water, $ABCDEF$. The sides $ABFE$ and $CDEF$ are rectangles. The triangular ends, $\angle ADE$ and $\angle BCF$ are isosceles and $AED = BFC = 90°$. The ends ADE and BCF are vertical and EF is horizontal. Given that $AD = x$ metres,

  a show that the area of $\triangle ADE$ is $\dfrac{1}{4}x^2 \text{ m}^2$.

Given also that the capacity of the container is 4000 m^3 and that the total area of the two triangular and two rectangular sides of the container is $S \text{ m}^2$,

  b show that $S = \dfrac{x^2}{2} + \dfrac{16\,000\sqrt{2}}{x}$.

Given that x can vary,

  c use calculus to find the minimum value of S,

  d justify that the value of S you have found is a minimum.   [Edexcel June 1998]

7 Integration

Note:
The case when $n = -1$ is described in module C4.

Recall from C1:

$$\int x^n \, dx = \frac{1}{n+1}x^{n+1} + c \quad (n \neq -1)$$

Recall:
Raise the power of x by 1 and divide by the new power.

If a is a constant:

$$\int ax^n \, dx = \frac{a}{n+1}x^{n+1} + c \quad (n \neq -1)$$

Tip:
Remember to include the integration constant, c.

Recall also that

$$\int (f(x) \pm g(x)) \, dx = \int f(x) \, dx \pm \int g(x) \, dx$$

Note:
This means that you integrate term by term.

7.1 Definite integrals

Evaluation of definite integrals.

A **definite integral** has the form $\int_a^b f(x)dx$, where a and b are the **limits** of integration.

Note:
a is the lower limit and b is the upper limit.

To **evaluate** a definite integral:

- integrate $f(x)$ but omit the integration constant, c
- substitute the upper limit
- subtract from this the value obtained when the lower limit is substituted.

Note:
You have to work out a specific value rather than give it as a function of x. It does not contain an integration constant c.

Example 7.1 Evaluate $\int_3^4 2x \, dx$.

Step 1: Integrate the function, simplifying if possible.

Step 2: Substitute the limits, upper limit first.

Step 3: Calculate the numerical value.

$$\int_3^4 2x \, dx = \left[\frac{2}{2}x^2\right]_3^4$$

$$= \left[x^2\right]_3^4 = 4^2 - 3^2$$

$$= 16 - 9 = 7$$

Tip:
When you have integrated, use square brackets and put the limits on the right.

Note:
The answer is a number, not a function of x.

Example 7.2 Evaluate $\int_{-3}^{-2} (5 - 4x)dx$.

Step 1: Integrate the function, simplifying if possible.

Step 2: Substitute the limits, upper limit first.

Step 3: Calculate the numerical value.

$$\int_{-3}^{-2} (5 - 4x)dx = \left[5x - \frac{4}{2}x^2\right]_{-3}^{-2}$$

$$= \left[5x - 2x^2\right]_{-3}^{-2}$$

$$= 5(-2) - 2(-2)^2 - (5(-3) - 2(-3)^2)$$

$$= -10 - 8 - (-15 - 18)$$

$$= -18 - (-33) = 15$$

Note:
Take great care with the negative numbers.

Example 7.3 Evaluate $\int_1^2 \frac{3}{x^4}\,dx$.

Recall:
Definite integration.

Step 1: Write term(s) in the form ax^n using the index laws.

$$\int_1^2 \frac{3}{x^4}\,dx = \int_1^2 3x^{-4}\,dx$$

Step 2: Integrate.

$$= \left[\frac{3}{3}x^{-3}\right]_1^2$$

$$= \left[-x^{-3}\right]_1^2$$

Step 3: Substitute the limits, upper limit first.

$$= (-2^{-3}) - (-1^3)$$

$$= -\tfrac{1}{8} + 1$$

$$= \tfrac{7}{8}$$

Tip:
Remember to substitute the upper limit first.

7.2 The area under a curve

Interpretation of the definite integral as the area under a curve.

Area under a curve

Integration can be used to find areas bounded by lines and curves. This is often referred to as finding the area 'under' a curve.

Consider the area of a region bounded by a curve $y = f(x)$, the x-axis and the lines $x = a$ and $x = b$.

If the area is *above* the x-axis:

$$\text{Area} = \int_a^b y\,dx$$

If the area is *below* the x-axis, the value of $\int_a^b y\,dx$ is negative.

The area is found by taking the positive value of the number calculated.

$$\text{Area} = \left|\int_a^b y\,dx\right|$$

Note:
The straight lines mean that you take the positive value. This is called the modulus.

Example 7.4 Find the area of the region enclosed by the curve $y = x^2 + 2$, the x-axis and the lines $x = 1$ and $x = 2$.

Step 1: Use the area formula with $a = 1$ and $b = 2$.

Step 2: Integrate term by term.

Step 3: Substitute the limits and evaluate.

$$\int_a^b y \, dx = \int_1^2 (x^2 + 2) \, dx$$

$$= \left[\tfrac{1}{3}x^3 + 2x \right]_1^2$$

$$= \tfrac{1}{3} \times 2^3 + 2 \times 2 - (\tfrac{1}{3} \times 1^3 + 2 \times 1)$$

$$= \tfrac{8}{3} + 4 - (\tfrac{1}{3} + 2)$$

$$= 2\tfrac{2}{3} + 4 - 2\tfrac{1}{3}$$

$$= 4\tfrac{1}{3}$$

The area is $4\tfrac{1}{3}$ units^2.

Note:
The area is above the curve, so the integral will give a positive number.

Tip:
Be careful with the fractions. It is a good idea to write down all the working.

Example 7.5 The diagram shows a sketch of the curve $y = x(x - 3)$.

Find the area enclosed between the curve and the x-axis.

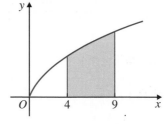

$y = x(x - 3)$

Tip:
Notice that the area is below the x-axis.

Step 1: Use the area formula with $a = 0$ and $b = 3$.

Step 2: Expand the brackets and integrate term by term.

Step 3: Substitute the limits and evaluate.

$$\int_a^b y \, dx = \int_0^3 x(x - 3) \, dx$$

$$= \int_0^3 (x^2 - 3x) \, dx$$

$$= \left[\tfrac{1}{3}x^3 - \tfrac{3}{2}x^2 \right]_0^3$$

$$= \tfrac{1}{3} \times 3^3 - \tfrac{3}{2} \times 3^2 - (0 - 0)$$

$$= 9 - 13\tfrac{1}{2}$$

$$= -4\tfrac{1}{2}$$

Tip:
Write the expression in index form before you integrate.

Tip:
It is helpful to put in the zeros to show that you have substituted for the lower limit.

Step 4: Find the modulus. $\quad \text{Area} = \left| \int_a^b y \, dx \right| = \left| -4\tfrac{1}{2} \right| = 4\tfrac{1}{2}$ units^2.

Note:
As expected, the value of the integral is negative as the area lies below the x-axis.

Example 7.6 The diagram shows the curve $y = \sqrt{x}$. Calculate the area of the shaded region.

Step 1: Use the area formula with $a = 4$ and $b = 9$

Step 2: Integrate.

Step 3: Substitute the limits and evaluate.

$$\int_a^b y \, dx = \int_4^9 \sqrt{x} \, dx$$

$$= \int_4^9 x^{\frac{1}{2}} \, dx$$

$$= \left[\frac{1}{\frac{3}{2}} x^{\frac{3}{2}} \right]_4^9$$

$$= \left[\tfrac{2}{3} x^{\frac{3}{2}} \right]_4^9$$

$$= \tfrac{2}{3} (9^{\frac{3}{2}} - 4^{\frac{3}{2}})$$

$$= \tfrac{2}{3} (27 - 8)$$

$$= 12\tfrac{2}{3}$$

The area is $12\tfrac{2}{3}$ units^2.

Area enclosed between a line and a curve

Example 7.7 The curve $y = x^2$ and the line $y = 2 - x$ intersect at $A(-2, 4)$ and $B(1, 1)$. Find the area enclosed between the line and the curve.

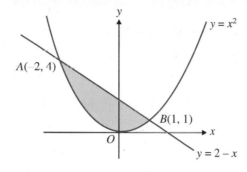

Step 1: Find the area under the line from $x = -2$ to $x = 1$.

First find the area under the line between $x = -2$ and $x = 1$.

Note:
You could evaluate
$$\int_{-2}^{1} (2 - x)dx$$
to find the area under the line.

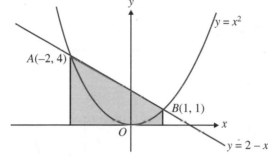

$$\text{Area of trapezium} = \tfrac{1}{2}(4 + 1) \times 3$$
$$= 7\tfrac{1}{2} \text{ units}^2$$

Recall:
Area of a trapezium $= \tfrac{1}{2}$ (sum of parallel sides) $\times$ perpendicular distance between them.

Now find the area under the curve between $x = -2$ and $x = 1$.

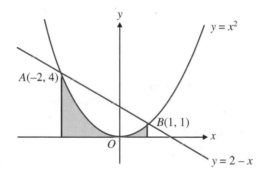

Step 2: Find the area under the curve from $x = -2$ to $x = 1$.

Area under the curve:
$$\int_{a}^{b} y \, dx = \int_{-2}^{1} x^2 dx$$
$$= \left[\tfrac{1}{3}x^3 \right]_{-2}^{1}$$
$$= \tfrac{1}{3}(1^3 - (-2)^3)$$
$$= \tfrac{1}{3} \times 9$$
$$= 3$$

Area under curve $= 3$ units^2

Step 3: Subtract the area under the curve from the area under the line.

Required area $= 7\tfrac{1}{2} - 3$
$$= 4\tfrac{1}{2} \text{ units}^2$$

Example 7.8 The curve $y = \dfrac{1}{x^2}$ and the line $y = x$ intersect at P as shown.

a Find the coordinates of P. **b** Calculate the area of the shaded region.

Step 1: Solve the simultaneous equations. **a** At P, $y = \dfrac{1}{x^2}$ and $y = x$. Solving the equations simultaneously gives

$$x = \frac{1}{x^2}$$

$(\times\ x^2)\quad x^3 = 1$

$$x = 1$$

Substituting into $y = x$ gives $y = 1$, so P is the point $(1, 1)$.

TIP:
You can substitute into either equation.

Step 2: Find the area under the line from $x = 0$ to $x = 1$. **b** Area of region between $y = x$, $x = 1$ and x-axis:

$\text{Area}_1 = \text{Area of triangle}$

$\qquad = \dfrac{1 \times 1}{2}$

$\qquad = \tfrac{1}{2}$

TIP:
You could evaluate $\displaystyle\int_0^1 y\,dx$ where $y = x$.

Step 3: Find the area under the curve from $x = 1$ to $x = 2$. Area of region between $y = \dfrac{1}{x^2}$, $x = 1$ and $x = 2$:

$\text{Area}_2 = \displaystyle\int_1^2 x^{-2}\,dx$

$\qquad = \left[\dfrac{1}{-1}x^{-1}\right]_1^2$

$\qquad = \left[-\dfrac{1}{x}\right]_1^2$

$\qquad = -\tfrac{1}{2} - (-1)$

$\qquad = \tfrac{1}{2}$

TIP:
Remember to write terms in index form before integrating.

Step 4: Add the areas. Total area $= \tfrac{1}{2} + \tfrac{1}{2} = 1\ \text{unit}^2$

Example 7.9 The curve $y = 1 + 2\sqrt{x}$ and the line $y = x + 1$ intersect at $P(0, 1)$ and $Q(4, 5)$. Find the area of the region enclosed between the line and the curve, shown shaded in the diagram.

Step 1: Find the area under the curve from $x = 0$ to $x = 4$. Let the area under the curve be A_1.

$\text{Area } A_1 = \displaystyle\int_0^4 (1 + 2\sqrt{x})\,dx$

$\qquad = \displaystyle\int_0^4 (1 + 2x^{\frac{1}{2}})\,dx$

$\qquad = \left[x + \dfrac{2}{\frac{3}{2}}x^{\frac{3}{2}}\right]_0^4$

$\qquad = \left[x + \tfrac{4}{3}x^{\frac{3}{2}}\right]_0^4$

$\qquad = 4 + \tfrac{32}{3} - 0$

$\qquad = 14\tfrac{2}{3}\ \text{units}^2$

Step 2: Find the area under the line from $x = 0$ to $x = 4$. Let the area under the line be A_2.

$\text{Area } A_2 = \dfrac{1 + 5}{2} \times 4 = 12\ \text{units}^2$

Step 3: Subtract to find the required area. Required area $= 14\tfrac{2}{3} - 12 = 2\tfrac{2}{3}\ \text{units}^2$

TIP:
Area of a trapezium
$= \dfrac{a + b}{2} \times h$

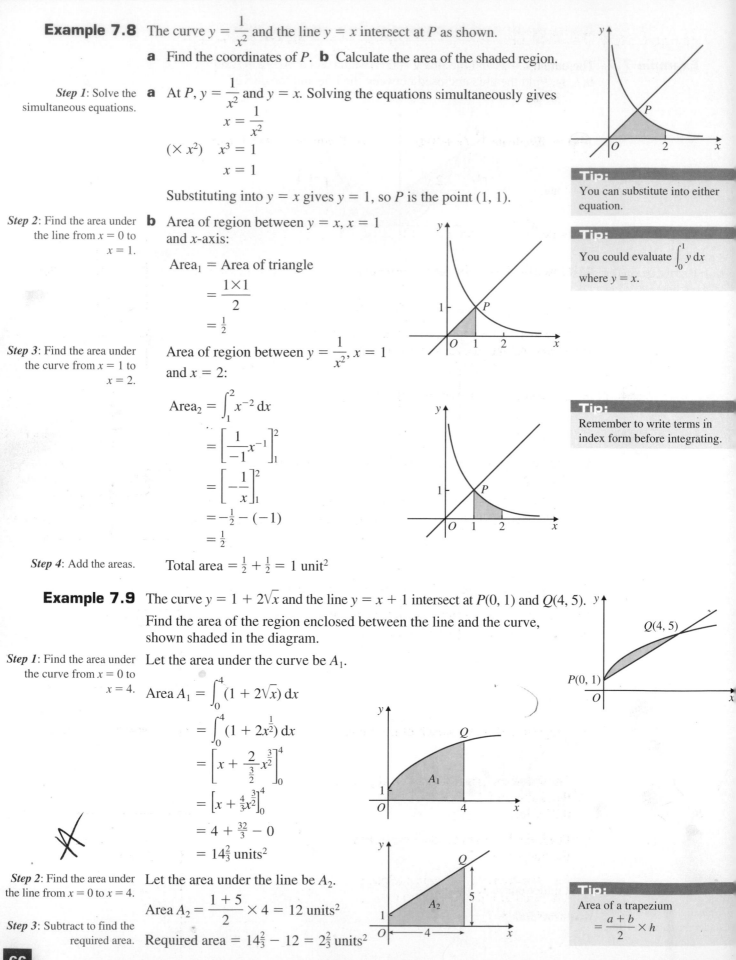

1 Evaluate the following definite integrals:

    a $\displaystyle\int_1^2 3x^2\,dx$
    b $\displaystyle\int_0^2 x^4\,dx$
    c $\displaystyle\int_2^3 \tfrac{1}{2}x\,dx$

2 ◎ **a** Evaluate $\displaystyle\int_{-2}^2 (x+3)\,dx$.
    b Evaluate $\displaystyle\int_{-2}^{-1} (x^3+x)\,dx$.

◎ **3** Evaluate   **a** $\displaystyle\int_1^2 \left(x-\frac{2}{x^2}\right)dx$
    b $\displaystyle\int_{-2}^{-1} \frac{1}{x^3}\,dx$.

4 a Write $\dfrac{x^3+1}{x^2}$ in the form $x^p + x^q$, where p and q are integers.

    b Hence find the value of $\displaystyle\int_1^2 \frac{x^3+1}{x^2}\,dx$.

5 The diagram shows a sketch of the curve $y = 4x^3 + 1$.

    Find the area of the region enclosed by the curve, the x-axis and the lines $x = 1$ and $x = 2$.

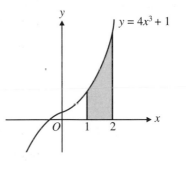

6 a Solve the simultaneous equations:
$$y = x^2 + 2$$
$$x + y = 4$$

    b The curve $y = x^2 + 2$ and the line $x + y = 4$ intersect at P and Q.

      i Write down the coordinates of P and Q.

      ii Find the area enclosed between the line and the curve.

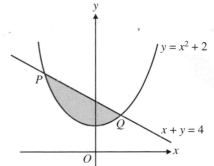

7 The diagram shows the graph of $y = -\dfrac{1}{x^3}$, for $x > 0$.

    Find the area of the region enclosed between the curve, the x-axis and the lines $x = 1$ and $x = 2$.

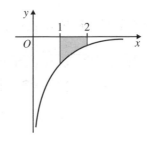

8 a Diagram 1 shows a sketch of the curve

$$y = \sqrt[3]{x}, \text{ for } x \geqslant 0.$$

    Show that the area enclosed by the curve, the x-axis and the line $x = 1$ is 0.75 units.^{2}

    b Diagram 2 shows the same curve and the line $y = x$.

    Find the area of the region enclosed between the line and the curve when $x \geqslant 0$.

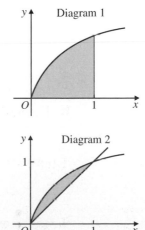

Diagram 1

Diagram 2

9 The diagram shows a sketch of the curve $y = 1 + \dfrac{1}{x^2}$ for $x > 0$.

The line $y = 2$ intersects the curve at $(1, 2)$.

Find the area of the region enclosed by the curve and the lines $x = 2$ and $y = 2$.

SKILLS CHECK **7A EXTRA** is on the CD

7.3 The trapezium rule

Approximation of area under a curve using the trapezium rule.

Consider the region enclosed by the curve $y = f(x)$, the x-axis and the lines $x = a$ and $x = b$. To find an approximation of the area of this region, split it into n strips, of equal width h, where $h = \dfrac{b - a}{n}$. Form trapezia by joining the top ends of each strip with a straight line.

Note:
Some trapezia give an overestimate.

Some give an underestimate.

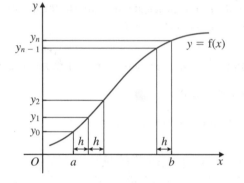

Note:
The y-value for each corresponding x-value is calculated. These y-values y_0, y_1, ..., y_n are called **ordinates**.

The area under the curve is given by $\displaystyle\int_a^b y \, dx$. An approximate value can be found using the **trapezium rule** with n intervals:

$$\int_a^b y \, dx \approx \tfrac{1}{2} h \left[(y_0 + y_n) + 2(y_1 + y_2 + \cdots + y_{n-1}) \right]$$

Tip:
Area of a trapezium $= \dfrac{h}{2}(a + b)$

Note:
You will be given this formula in the examination.

Example 7.10 The diagram shows the region bounded by the curve $y = \sqrt{x - 1}$, the x-axis and the lines $x = 2$ and $x = 6$.

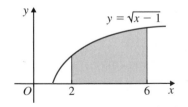

a Find an approximation to the area of the region, using the trapezium rule with four strips (five ordinates).

b Is your value an overestimate or an underestimate?

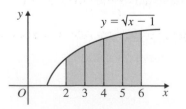

Note:
The number of strips is the same as the number of intervals. The number of ordinates is the same as the number of lines. Four strips are equivalent to five ordinates.

Step 1: Find the number of intervals, n, and the height h, where $h = \dfrac{b-a}{n}$.

Step 2: Fill in the table shown.

a $n = 4, h = \dfrac{6-2}{4} = 1. \; x = 2, 3, 4, 5, 6.$

x	2	3	4	5	6
$y = \sqrt{x-1}$	$\sqrt{1}$	$\sqrt{2}$	$\sqrt{3}$	$\sqrt{4}$	$\sqrt{5}$
y_n	y_0	y_1	y_2	y_3	y_4

Note:
x-values go from lower limit to upper limit in jumps of h. The y-values are calculated for each x-value using the function under the integral.

Step 3: Substitute values into the trapezium rule.

$$\int_a^b y \, dx \approx \tfrac{1}{2} \times 1 \, [(\sqrt{1} + \sqrt{5}) + 2(\sqrt{2} + \sqrt{3} + \sqrt{4})]$$
$$= 6.764\ldots$$
$$= 6.76 \text{ (3 s.f.)}$$

Tip:
To avoid rounding errors, it is better not to calculate the values until the final line.

Step 4: Decide whether the value is an overestimate or underestimate.

b From the graph, the value calculated by the trapezium rule is an underestimate for the area of the region.

Note:
It can be shown by integration (in C4 module) that the area is 6.79 to 3 s.f.

Note on graphical calculators

Some graphical calculators can be programmed to find numerical approximations of definite integrals. These provide a useful check, but you will not be awarded marks for a question on the trapezium rule unless appropriate working has been shown.

Example 7.11 Find an approximation to $\displaystyle\int_0^1 \sqrt{\sin x} \, dx$, where x is in radians, using the trapezium rule with six ordinates.

Step 1: Find the number of intervals n and the height h.

$$h = \dfrac{1-0}{5} = 0.2; \; x = 0, 0.2, 0.4, \ldots, 1.$$

Step 2: Fill in the table shown.

x	0	0.2	0.4	0.6	0.8	1
$y = \sqrt{\sin x}$	0	$\sqrt{\sin 0.2}$	$\sqrt{\sin 0.4}$	$\sqrt{\sin 0.6}$	$\sqrt{\sin 0.8}$	$\sqrt{\sin 1}$
y_n	y_0	y_1	y_2	y_3	y_4	y_5

Tip:
It is a good idea to write your working in a table so that you can see the values clearly.

Step 3: Substitute values into the trapezium rule.

$$\int_0^1 \sqrt{\sin x} \, dx \approx \tfrac{1}{2} \times 0.2 \, [(0 + \sqrt{\sin 1}) + 2(\sqrt{\sin 0.2} + \sqrt{\sin 0.4}$$
$$+ \sqrt{\sin 0.6} + \sqrt{\sin 0.8})]$$
$$= 0.6253\ldots$$
$$= 0.625 \text{ (3 s.f.)}$$

Tip:
Remember to set your calculator to radian mode.

SKILLS CHECK **7B: The trapezium rule**

1 **a** Sketch the graph of $y = 2^x$.

   b Estimate $\displaystyle\int_0^4 2^x \, dx$, using the trapezium rule with four intervals.

   c State whether your estimate is an overestimate or an underestimate.

2 The diagram shows a sketch of $y = \dfrac{1}{1+x}$ for $x > -1$.

   Estimate $\displaystyle\int_1^2 \dfrac{1}{1+x} \, dx$ using the trapezium rule with six ordinates.

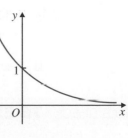

3 The following is a table of values, correct to three decimal places, for $y = \cos x$, where x is in radians.

x	0	0.2	0.4	0.6	0.8	1
y	1	0.980	0.921	p	0.697	q

a Find the value of p and the value of q.

b Use the trapezium rule and the values of y in the completed table to obtain an estimate for
$$\int_0^1 \cos x \, dx.$$

4 The diagram shows a sketch of $y = \log_{10} x$.

   a Divide the shaded region into five equal-width intervals.

   b Use the trapezium rule to estimate $\int_1^3 \log_{10} x \, dx$.

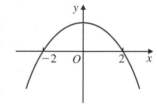

5 a Sketch the graph of $y = 10^{-x}$.

   b Estimate $\int_{-2}^{-1} 10^{-x} \, dx$, using the trapezium rule with five strips.

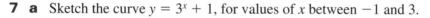 **6** The diagram shows a sketch of $y = 4 - x^2$.

   a Estimate the area of the region enclosed between the curve and the x-axis, using the trapezium rule with four strips.

   b Evaluate the area exactly using integration and calculate the percentage error in taking your answer in **a** as the area.

7 a Sketch the curve $y = 3^x + 1$, for values of x between -1 and 3.

   b Use the trapezium rule with four strips to estimate the area of the region bounded by the curve, the x-axis and the lines $x = -1$ and $x = 3$.

   c Explain briefly how the trapezium rule could be used to find a more accurate estimate of the area of the required region.

8 The values of a function $f(x)$ are given in the table.

x	1	2	3	4	5	6
$f(x)$	3.5	5	7.5	11	15.5	21

Find an approximate value, using the trapezium rule with five strips, for

a $\int_1^6 f(x) \, dx$ **b** $\int_1^6 \dfrac{1}{f(x)} \, dx$.

9 a Tabulate, correct to two decimal places, the values of the function $f(x) = \dfrac{2}{2 + x^2}$ for values of x from 0 to 2 at intervals of 0.4.

   b Use the values found in part **a** to estimate $\int_0^2 \dfrac{2}{2 + x^2} \, dx$.

1 i Differentiate with respect to x $2x^3 + \sqrt{x} + \dfrac{x^2 + 2x}{x^2}$.

   ii Evaluate $\displaystyle\int_1^4 \left(\dfrac{x}{2} + \dfrac{1}{x^2}\right) dx.$ [Edexcel Nov 2002]

2 The diagram shows the line with equation $y = 9 - x$ and the curve with equation $y = x^2 - 2x + 3$. The line and the curve intersect at the points A and B, and O is the origin.

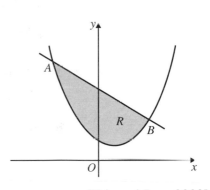

   a Calculate the coordinates of A and the coordinates of B.

The shaded region R is bounded by the line and the curve.

   b Calculate the area of R.

[Edexcel June 2003]

3 The diagram shows the line with equation $y = x + 1$ and the curve with equation $y = 6x - x^2 - 3$.

The line and the curve intersect at the points A and B, and O is the origin.

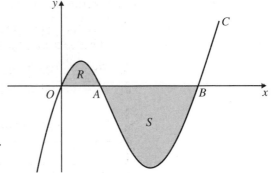

   a Calculate the coordinates of A and the coordinates of B.

The shaded region R is bounded by the line and the curve.

   b Calculate the area of R.   [Edexcel Jan 2002]

4 The diagram shows part of the curve C with equation $y = f(x)$, where

$$f(x) = x^3 - 6x^2 + 5x.$$

The curve crosses the x-axis at the origin O and at the points A and B.

   a Factorise $f(x)$ completely.

   b Write down the x-coordinates of the points A and B.

   c Find the gradient of C at A.

The region R is bounded by C and the line OA, and the region S is bounded by C and the line AB.

   d Use integration to find the area of the combined regions R and S, shown shaded in the diagram.

[Edexcel Nov 2003]

5 The diagram shows a sketch of $y = 2^{-x}$.

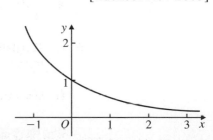

   a Use the trapezium rule with six ordinates (five strips) to find an approximation for $\displaystyle\int_0^2 2^{-x}\, dx$.

   b By considering the graph of $y = 2^{-x}$, explain with the aid of a diagram whether your approximation will be an overestimate or an underestimate of the true value of $\displaystyle\int_0^2 2^{-x}\, dx$.

6 A measure of the effective voltage, M volts, in an electrical circuit is given by

$$M^2 = \int_0^1 V^2 \, dt$$

where V volts is the voltage at time t seconds. Pairs of values of V and t are given in the following table.

t	0	0.25	0.5	0.75	1
V	-48	207	37	-161	-29
V^2					

Use the trapezium rule with five values of V^2 to estimate the value of M. [Edexcel June 2001]

7 The diagram shows part of the curve with equation

$$y = x^3 - 6x^2 + 9x.$$

The curve touches the x-axis at A and has a maximum turning point at B.

a Show that the equation of the curve may be written as

$$y = x(x - 3)^2,$$

and hence write down the coordinates of A.

b Find the coordinates of B.

The shaded region R is bounded by the curve and the x-axis.

c Find the area of R. [Edexcel June 2001]

8 The diagram shows part of the curve C with equation

$$y = \tfrac{3}{2}x^2 - \tfrac{1}{4}x^3.$$

The curve C touches the x-axis at the origin and passes through the point $A(p, 0)$.

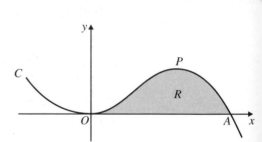

a Show that $p = 6$.

b Find an equation of the tangent to C at A.

The curve C has a maximum at the point P.

c Find the x-coordinate of P.

The shaded region R, in the diagram, is bounded by C and the x-axis.

d Find the area of R. [Edexcel P1 Jan 2004]

9 The diagram shows a sketch of $y = \dfrac{1}{1 + x^2}$.

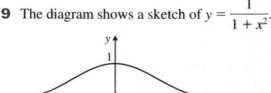

Use the trapezium rule with five ordinates (four strips) to find an approximation for $\displaystyle\int_{-2}^{2} \frac{1}{1 + x^2} \, dx.$

Practice exam paper

Answer **all** questions.

Time allowed: 1 hour 30 minutes

A calculator is **allowed** in this paper.

1 $f(x) = x^3 + ax + b$

Given that $(x - 2)$ is a factor of $f(x)$ and that, when $f(x)$ is divided by $(x + 1)$, the remainder is 6, find the value of the constant a and the value of the constant b. *(4 marks)*

2

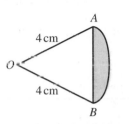

The diagram shows a sector AOB of a circle, centre O and radius 4 cm, and a shaded segment of the circle. Given that the area of the sector is 2.8 cm^2, calculate

a the size, in radians, of angle AOB, *(2 marks)*

b the area, to 3 significant figures, of triangle AOB, *(2 marks)*

c the area, of the shaded segment. *(2 marks)*

3 Solve $\log_9 (x^2 + 4x - 9) = \frac{1}{2}$. *(6 marks)*

4 $f(x) = \left(1 + \dfrac{x}{n}\right)^7$.

Given that, in the binomial expansion of $f(x)$ in terms of x, the coefficient of x is equal to the coefficient of x^2,

a calculate the value of n. *(4 marks)*

Using the value of n found in part **a**,

b calculate the value of the coefficient of x^4 in the binomial expansion of $f(x)$, giving your answer as an exact fraction. *(3 marks)*

5 An estimate of $\displaystyle\int_0^2 \sqrt{(6 - x - x^2)} \, dx$ is to be made using the trapezium rule and the table below.

x	0	0.5	1	1.5	2
y	2.45	2.29	p	q	r

a Write down the values of p, q and r. *(2 marks)*

b Using the trapezium rule and all of the values in the table, find an estimate of the integral, giving your answer to 3 significant figures. *(4 marks)*

The accurate value of the integral, to 3 significant figures, is 3.67.

c Calculate, to 2 significant figures, the percentage error of the answer to part **b**. *(2 marks)*

6 The diagram shows the graph of part of the curve C with equation $y = 7x - 2x^2$ and part of the line l with equation $y = x$. The curve C and the line l intersect at the points O and A.

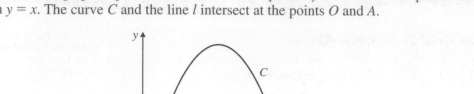

a Find the coordinates of A. (2 marks)

The finite region R is bounded by C and l.

b Use integration to find the area of R. (7 marks)

7 The circle C, with centre X, has equation

$$x^2 + y^2 - kx + 6y = 0.$$

The point P, with coordinates $(7, 1)$, lies on C.

a Calculate the value of the constant k. (2 marks)

b Find the coordinates of X. (2 marks)

c Find the equation of the tangent to C at P, giving your answer in the form $ax + by = c$, where a, b and c are integers. (5 marks)

8 An open cylindrical can, made of thin material, has the shape of a cylinder closed at one end and open at the other end. The radius of the circular cross-section of the can is r cm and the height of the can is h cm. The total surface area of the can, including the base, is A cm^2. The volume of the can is 64π cm^3.

a Show that $A = \pi\left(r^2 + \dfrac{128}{r}\right)$. (3 marks)

b Use calculus to find the value of r for which A has a stationary value. (4 marks)

c Find the stationary value of A, establishing whether this is a maximum value or a minimum value. (4 marks)

9 a The sum of the first and second term of a geometric series is 4 and the sum of the first and third term is 25.

    i Calculate the possible values of the common ratio of the series. (5 marks)

    Given also that the series has a sum to infinity S,

    ii calculate the value of S. (4 marks)

b The first three terms of another geometric series are

$$2\sin\theta, \ 3 \text{ and } \frac{2}{\cos\theta},$$

where $-\pi < \theta \leqslant \pi$.

    i Show that $\tan\theta = \dfrac{9}{4}$. (3 marks)

    ii Calculate, to 2 decimal places, the values of θ. (3 marks)

Answers

SKILLS CHECK 1A (page 4)

1 a $\frac{1}{2}, 2, -\frac{1}{3}$ **b** $\pm 2, \pm 3$ **c** $\frac{1}{4}, -\frac{4}{3}, 2$ **d** $-3, 1$

2 a $1, 2, -3$ **b** $3, -2 \pm \sqrt{2}$ **c** $-1, -2, -3$

3 a -1 **b** 8

   c -1 **d** $0, (x-1)$ is a factor

4 a $x^2 - 3x - 5 - \dfrac{18}{x-3}$ **b** $x^2 + 5x - 6 + \dfrac{7}{x+1}$

   c $2x^2 + 5x + 6 + \dfrac{17}{x-2}$

5 $2, 2 \pm \sqrt{11}$

6 $(x+1)^3$

7 $a = 4, b = -1$

Exam practice 1 (page 5)

1 -4.75

2 b $(x+3)(x-4)(x-2)$

3 a $a = -2, b = 5$

4 a $p = -20$ **b** 1

5 b $(x+3)(x-5)(x+2)$

6 a 0 **b** 0

   c $(x+3)(x-1)(2x-1)$

7 a -19 **c** $(x+3)(3x-1)(x-2)$

8 b $(x-5)$ **c** $(x+7)(x+6)(x-5)$

   d $-7, -6, 5$

9 a $a = -1, b = 2$

SKILLS CHECK 2A (page 7)

1 a $(x-3)^2 + (y+2)^2 = 16$ **b** $(x+5)^2 + y^2 = 25$

2 a $(x-1)^2 + (y-2)^2 = 5^2, (1, 2), 5$

   b $(x+5)^2 + (y+12)^2 = 13^2, (-5, -12), 13$

   c $(x-3)^2 + (y+5)^2 = 4^2, (3, -5), 4$

   d $(x+1)^2 + (y-3)^2 = 7, (-1, 3), \sqrt{7}$

3 $(x-2)^2 + (y+5)^2 = 5^2$

4 $(x-1)^2 + (y+4)^2 = 10$

5 $(1, 1), \sqrt{5}; (7, 4), 2\sqrt{5}$

6 a $-\dfrac{2}{3}$ **b** $y = \dfrac{3}{2}x - \dfrac{11}{2}$

7 a $(x-2)^2 + (y+3)^2 = 4^2$ **b** $4, (2, -3)$ **c** $\begin{pmatrix} 2 \\ -3 \end{pmatrix}$

8 $x^2 + y^2 - 10x + 6y - 15 = 0$

SKILLS CHECK 2B (page 12)

1 c $(0, 4), \sqrt{17}$ **d** $x^2 + (y-4)^2 = 17$

2 $k = 4, -1$

3 b Yes. C on perp. bisector of AB

4 a $(1, 2)$

5 a $y = 2x - 2$ **b** $y = -x + 4$ **c** $(2, 2)$

6 $y = \dfrac{1}{3}x - \dfrac{10}{3}$

SKILLS CHECK 2C (page 13)

1 $y = -x + 4$

2 $y = \dfrac{1}{2}x + \dfrac{3}{2}$

3 $y = -\dfrac{1}{3}x + \dfrac{8}{3}$

4 $y = 4x - 3, y = -\dfrac{1}{4}x - 3; (0, -3)$

5 $y = \dfrac{1}{2}x + 2, y = 2x - 6$

6 a $(2, -5), 5$ **c** $3x - 4y - 26 = 0$

SKILLS CHECK 2D (page 15)

1 a $(3, 4), (4, 3)$ **b** $(1, 3), (4, 6)$ **c** $(-2, -1)$

2 $x = 1, y = -2$

3 a $x = 0, y = 0; x = 6, y = 0$. The circle intersects the x-axis at $(0, 0)$ and $(6, 0)$.

   b None. The line and circle do not intersect.

4 b $(-3, 4), (-4, 3)$

5 a $2x + 3y = 17$ **b** $P(1, 5)$ **c** $(x-4)^2 + (y-3)^2 = 13$

Exam practice 2 (page 15)

1 a $(5, -3)$ **b** 7

2 a centre $(3, -4)$ radius 10 **b** $(9, 4)$

3 a $(x-3)^2 + (y-4)^2 = 18$ **b** $(2 \pm 2\sqrt{2}, 5 \pm 2\sqrt{2})$ **c** 8

4 a $(x-1)^2 + (y-4)^2 = 9$ **b** inside

5 a $(2.5, 1)$ **b** $6x - 11y - 4 = 0$

   c $11x + 6y - 112 = 0$ **d** $\left(\dfrac{112}{15}, \dfrac{224}{45} \right)$

6 a $(2, -3), 5$ **b** $(5, 1)$ **c** $\frac{25}{6}$

7 a $(1, -3), 5$ **b** $5\sqrt{2}$ units

8 a i $(0, 5)$ **ii** 3

   c $x = 2.4$

   d L only intersects with C once, so L must be a tangent to C.

SKILLS CHECK 3A (page 20)

1 a i $31\,250$ **ii** $39\,062$

   b i $\frac{7}{64}$ **ii** $13\frac{57}{64}$ **iii** 14

   c i 64 **ii** 43

2 $-59\,048$

3 £123\,000

4 a $a = 40, r = \frac{1}{4}$ **b** $53\frac{1}{3}$

5 a $-\frac{1}{3}$ **b** $\frac{1}{324}$

6 a $699\,048$ **b** 3

7 $137\,262$

8 a £512   **b** £1023

9 a 16 **b** $p = 14, q = 6$

10 1

SKILLS CHECK 3B (page 25)

1 $16 - 96x + 216x^2 - 216x^3 + 81x^4$

2 $1 + 28y + 336y^2 + 2240y^3$

3 a $243 - 810x + 1080x^2$ **b** $A = 1215, B = -3564, C = 3780$

4 a $1 + 12x + 60x^2 + 160x^3$ **b** 1.1262 (4 d.p.)

5 $0.972\,333\,8$ (7 d.p.)

6 $40\sqrt{6}$

7 4

8 a $n = 8, k = -\frac{1}{2}$ **b** $-7x^3$

9 a $1 + 6ax + 15a^2x^2$ **b** $a = 3, b = 2$

10 a $1 - 15x + 90x^2 - 270x^3$ **b** -180

Exam practice 3 (page 26)

1 a $r = 0.8$ **b** $a = 10$

   c 50 **d** 0.189 (3 d.p.)

2 a $r = 0.4, a = 200$ **b** $333\frac{1}{3}$

   c 8.948×10^{-4}

3 a $512 + 576x + 288x^2 + 84x^3$ **b** 572.564

4 a $1 + \binom{n}{1}ax + \binom{n}{2}(ax)^2 + \binom{n}{3}(ax)^3 + \cdots$

   b $16, \frac{1}{2}$ **c** 70

5 a $510\,000, 520\,500, 531\,525$. Net growth, more fish produced than caught

   c $25\,000$

6 c $1 + 7x + \dfrac{91}{4}x^2 + \dfrac{91}{2}x^3 + \dfrac{1001}{16}x^4 + \dfrac{1001}{16}x^5 + \cdots$

7 $9, -2$

8 $\frac{3}{4}, -144$

9 a $1 + \binom{n}{1}3x + \binom{n}{2}(3x)^2 + \binom{n}{3}(3x)^3 + \cdots$

   b 12

   c $40\,095$

SKILLS CHECK 4A (page 32)

1. 7.8 cm
2. 15.8 cm (3 s.f.)
3. **a** 24.5° (3 s.f.) **b** 11.0 cm² (3 s.f.)
4. **a** 52.1° (3 s.f.) **b** 92.9° (3 s.f.)
5. **a** 8.25 cm (3 s.f.) **b** 35.8 cm² (3 s.f.)
6. **b** **i** 11.6 cm² (3 s.f.), 17.7 cm² (3 s.f.)
   ii 4.10 cm (3 s.f.), 9.19 cm (3 s.f.)
7. 55.9° (3 s.f.)
8. **a** 19.2 cm (3 s.f.) **b** 69.2 cm² (3 s.f.)
9. **a** 17.9 m (3 s.f.) **b** 57.1° (3 s.f.)
   c 60.9° (3 s.f.) **d** 265 m² (3 s.f.)
10. **a** 30°, 6.46 cm (3 s.f.) **b** 19.3 cm (3 s.f.)
    c $\sin A = 0.5 \Rightarrow a = 30°$, 150° giving two triangles with same base and same perpendicular height (see CD for diagrams).

SKILLS CHECK 4B (page 35)

1. **a** 4.89 radians (3 s.f.) **b** 85.9° (3 s.f.)
2. **a** 120° **b** 135° **c** 270° **d** 105°
3. **a** $\frac{1}{4}\pi$ **b** $\frac{5}{6}\pi$ **c** $\frac{11}{6}\pi$ **d** $\frac{4}{3}\pi$
4. **a** 3 cm **b** 13 cm
   c 7.06 cm² (3 s.f.) **d** 7.5 cm²
5. **a** 1.2 radians **b** 64.9 cm² (3 s.f.)
6. **a** 1.5 radians **b** 21 cm
7. **a** 0.848 radians **b** 27.1 cm² (3 s.f.) **c** 3.14 cm² (3 s.f.)
8. **a** $6 + 17\theta$ **c** 12.75 cm²
9. **a** 4.84 cm (3 s.f.) **b** 1.05 cm (3 s.f.) **c** 16.9 cm (3 s.f.)
   d 13.5 cm² **e** 1.05 cm² (3 s.f.) **f** 12.5 cm² (3 s.f.)
10. **a** 4.57 cm (3 s.f.) **b** 1.34 cm² (3 s.f.)

SKILLS CHECK 4C (page 44)

1. **a** 17°, 163° (nearest °) **b** 60°, 300°
   c 124°, 304° (nearest °) **d** 105°, 165°, 285°, 345°
2. **a** $\frac{1}{3}\pi, \frac{2}{3}\pi$ **b** $\frac{3}{4}\pi$
   c $\frac{2}{3}\pi$ **d** $\frac{1}{9}\pi, \frac{5}{9}\pi, \frac{7}{9}\pi$
3. Curve for $y = \cos(x + 30°)$: translate $y = \cos x$ 30° to the left.
4. Curve for $y = \sin(x - \frac{1}{4}\pi)$: translate $y = \sin x$ $\frac{1}{4}\pi$ to the right.
5. $-90°, -30°, 30°, 90°$
6. $-360°, -315°, -225°, -180°, 0°, 45°, 135°, 180°, 360°$
8. $\frac{4}{3}$
9. **a** 1 **b** $\frac{1}{12}\pi, \frac{5}{12}\pi, \frac{3}{4}\pi$
10. $\frac{1}{2}\pi, \frac{11}{6}\pi$
11. **b** 0°, 120°, 240°, 360° **c** 0°, 60°, 120°, 180°

Exam practice 4 (page 45)

1. **a** 77 m (nearest m) **b** 874 m² (3 s.f.)
2. 16 hectares
3. **a** 60 cm **b** 98.6 cm (nearest cm)
4. **a** 225, 345 **b** 22.2, 67.8, 202.2, 247.8 (1 d.p.)
5. **b** 6.768 cm² **c** 15.682 cm² **d** 22.512 cm
6. **a** Sine graph translated 30 units to left on x-axis
   b (0, 0.5), (150, 0), (330, 0) **c** 180, 300
7. 0, 131.8, 228.2
8. **a** Graph is stretched sine curve with x-axis intercepts at 0, 60, 120, 180, and turning points at $y = 5, -5$.
   b max (30, 5), (150, 5); min (90, -5)
   c 10, 50, 130 170
9. $-160.6, -19.5, 90$
10. **a** Sketch of $y = \cos x$ graph moved to the left by $\frac{1}{4}\pi$. x-axis intercepts $(\frac{1}{4}\pi, 0)$, $(\frac{5}{4}\pi, 0)$; y-axis intercept $\left(0, \frac{\sqrt{2}}{2}\right)$. Turning points $(\frac{3}{4}\pi, -1)$ and $(\frac{7}{4}\pi, 1)$.
    b $(\frac{1}{4}\pi, 0)$, $(\frac{5}{4}\pi, 0)$, $\left(0, \frac{\sqrt{2}}{2}\right)$ **c** $x = \frac{1}{12}\pi, \frac{17}{12}\pi$
11. **i** $x = 46.6, 173.4$ **ii** **b** $\frac{2\sqrt{2}}{3}$

SKILLS CHECK 5A (page 51)

1. **a** 3 **b** 2 **c** 3 **d** $\frac{1}{2}$
2. **a** 9 **b** -2 **c** $\frac{3}{2}$ **d** $\frac{3}{2}$
3. **a** 2 **b** 36
4. **a** 5 **b** $-\frac{1}{2}$ **c** 3
5. $\log_2 p + \frac{1}{2}\log_2 q - \frac{1}{2} - \frac{3}{2}\log_2 r$
6. 36
7. **a** **i** 1 **ii** 3 **b** -3
8. **a** 4.75 (3 s.f.) **b** 0.945 (3 s.f.)
   c 2.92 (3 s.f.) **d** 1.43 (3 s.f.) **e** 1.77 (3 s.f.)
9. **a** $a = d = \log_a b$ **b** 55
10. **b** $x = 0, 2$
11. $x = 9, 27$
12. $p = 142, q = 70.2$

Exam practice 5 (page 51)

1. **a** Exponential curve which crosses the y-axis at (0, 1)
   b £1184 **c** 18
2. **a** $\frac{p}{4}$ **b** $1 + \frac{3p}{4}$
3. **a** $\frac{x+3}{x}$ **b** $\frac{1}{5}$
4. $x = \frac{3}{22}, y = 2\frac{2}{11}$
5. **b** 2.32
6. **a** -3 **b** -1 **c** $-3 - \frac{t}{2}$
7. **b** $\frac{1}{4}, \frac{3}{2}$ **d** 0.585
8. 7.1
9. **b** $x = \frac{1}{2}$
10. **a** $\frac{x+3}{x-1}$ **b** $-\frac{3}{4}$

SKILLS CHECK 6A (page 59)

1. **a** (0, 0) min, $\left(-\frac{2}{3}, \frac{4}{27}\right)$ max **b** (1, -2) min, (-1, 2) max
   c (-2, 26) max, (1, -1) min
   d (0, -3) min, (1, -2) stationary point of inflexion
2. $x = -\frac{1}{2}$ max, $x = \frac{2}{3}$ min
3. **a** 2 **b** $8\sqrt{2}$
4. **a** **i** $x > 2$ **ii** $x < 2$
   b **i** $x > 0$ **ii** $x < 0$
   c **i** $x > 1, x < -1$ **ii** $-1 < x < 1$
5. **a** $x > \frac{1}{12}$
6. 500 000 m²
7. 10 000
8. **b** 9 **c** 972 cm²
9. **c** 1728 cm²

Exam practice 6 (page 60)

1. **a** $7\frac{31}{32}$
2. **a** $v = 20$ **c** £30
3. **b** 5.42 **d** 277
4. **a** $2x + 4y + \pi x$ **b** $4xy + \frac{\pi x^2}{2}$
   d 14, 7 **e** 700
5. **a** 4 **b** (1, -9) min, $(-\frac{4}{3}, \frac{100}{27})$ max
6. **a** $4x^3 - 16x$ **b** (-2, -13), (0, 3), (2, -13)
   c min, max, min **d** $-x + 12y + 49 = 0$
7. **b** 36 **c** At $x = 3$, $\frac{\mathrm{d}^2V}{\mathrm{d}x^2} = -12 < 0 \Rightarrow V = 36$ is a maximum value
8. **c** 1200 **d** At $x = 28$, $\frac{\mathrm{d}^2S}{\mathrm{d}x^2} = -112 < 0 \Rightarrow S = 1200$ is a minimum value

SKILLS CHECK 7A (page 67)

1 a 7 **b** 6.4 **c** 1.25
2 a 12 **b** -5.25
3 a $\frac{1}{2}$ **b** $-\frac{3}{8}$
4 a $x + x^{-2}$ **b** 2
5 16 units^2
6 a $x = 1, y = 3; x = -2, y = 6$
  b i $P(-2, 6), Q(1, 3)$ **ii** $4\frac{1}{2}$ units^2
7 $\frac{3}{8}$ units^2
8 b 0.25 units^2
9 $\frac{1}{2}$ units^2

SKILLS CHECK 7B (page 69)

1 b 22.5
  c Underestimate
2 0.406 (3 s.f.)
3 a $p = 0.825, q = 0.540$ **b** 0.84 (2 d.p.)
4 a Strips have width 0.4 **b** 0.559 (3 s.f.)
5 b 39.8 (3 s.f.)
6 a 10 units^2 **b** $10\frac{2}{3}$ units^2, 6.25%
7 b $30\frac{2}{3}$ units^2 **c** Split into more strips
8 a 51.25 **b** 0.655 (3 s.f.)
9 a 1, 0.93, 0.76, 0.58, 0.44, 0.33 **b** 1.35 (3 s.f.)

Exam practice 7 (page 71)

1 a $6x^2 + \frac{1}{2}x^{-\frac{1}{2}} - 2x^{-2}$ **b** $4\frac{1}{2}$
2 a $A(-2, 11), B(3, 6)$ **b** $20\frac{5}{6}$
3 a $A(1, 2), B(4, 5)$ **b** $4\frac{1}{2}$
4 a $x(x - 1)(x - 5)$ **b** $x = 1$ for A, $x = 5$ for B
  c -4 **d** $\frac{131}{4}$
5 a 1.09 (3 s.f.) **b** Overestimate
6 133.9
7 a $(3, 0)$ **b** $(1, 4)$ **c** $\frac{27}{4}$
8 b $y = -9x + 54$ **c** 4 **d** 27
9 2.2

Practice exam paper (page 73)

1 $a = -5, b = 2$
2 a 0.35 **b** 2.74 cm^2 **c** 0.06 cm^2
3 $x = -6, 2$
4 a $n = 3$ **b** $\frac{35}{81}$
5 a $p = 2, q = 1.5, r = 0$ **b** 3.51 **c** 4.4%
6 a $(3, 3)$ **b** 9 units^2
7 a $k = 8$ **b** $(4, -3)$ **c** $3x + 4y = 25$
8 b $r = 4$ **c** $A = 48\pi$, the value is a minimum
9 a i $-\frac{3}{4}, 7$ **ii** $S = \frac{64}{7}$
  b ii $\theta \approx 1.15, -1.99$

SINGLE USER LICENCE AGREEMENT FOR CORE 2 FOR EDEXCEL CD-ROM
IMPORTANT: READ CAREFULLY

WARNING: BY OPENING THE PACKAGE YOU AGREE TO BE BOUND BY THE TERMS OF THE LICENCE AGREEMENT BELOW.

This is a legally binding agreement between You (the user or purchaser) and Pearson Education Limited. By retaining this licence, any software media or accompanying written materials or carrying out any of the permitted activities You agree to be bound by the terms of the licence agreement below.

If You do not agree to these terms then promptly return the entire publication (this licence and all software, written materials, packaging and any other components received with it) with Your sales receipt to Your supplier for a full refund.

YOU ARE PERMITTED TO:

- Use (load into temporary memory or permanent storage) a single copy of the software on only one computer at a time. If this computer is linked to a network then the software may only be used in a manner such that it is not accessible to other machines on the network.

- Transfer the software from one computer to another provided that you only use it on one computer at a time.

- Print a single copy of any PDF file from the CD-ROM for the sole use of the user.

YOU MAY NOT:

- Rent or lease the software or any part of the publication.

- Copy any part of the documentation, except where specifically indicated otherwise.

- Make copies of the software, other than for backup purposes.

- Reverse engineer, decompile or disassemble the software.

- Use the software on more than one computer at a time.

- Install the software on any networked computer in a way that could allow access to it from more than one machine on the network.

- Use the software in any way not specified above without the prior written consent of Pearson Education Limited.

- Print off multiple copies of any PDF file.

ONE COPY ONLY

This licence is for a single user copy of the software

PEARSON EDUCATION LIMITED RESERVES THE RIGHT TO TERMINATE THIS LICENCE BY WRITTEN NOTICE AND TO TAKE ACTION TO RECOVER ANY DAMAGES SUFFERED BY PEARSON EDUCATION LIMITED IF YOU BREACH ANY PROVISION OF THIS AGREEMENT.

Pearson Education Limited and/or its licensors own the software.
You only own the disk on which the software is supplied.

Pearson Education Limited warrants that the diskette or CD-ROM on which the software is supplied is free from defects in materials and workmanship under normal use for ninety (90) days from the date You receive it. This warranty is limited to You and is not transferable. Pearson Education Limited does not warrant that the functions of the software meet Your requirements or that the media is compatible with any computer system on which it is used or that the operation of the software will be unlimited or error free.

You assume responsibility for selecting the software to achieve Your intended results and for the installation of, the use of and the results obtained from the software. The entire liability of Pearson Education Limited and its suppliers and your only remedy shall be replacement free of charge of the components that do not meet this warranty.

This limited warranty is void if any damage has resulted from accident, abuse, misapplication, service or modification by someone other than Pearson Education Limited. In no event shall Pearson Education Limited or its suppliers be liable for any damages whatsoever arising out of installation of the software, even if advised of the possibility of such damages. Pearson Education Limited will not be liable for any loss or damage of any nature suffered by any party as a result of reliance upon or reproduction of or any errors in the content of the publication.

Pearson Education Limited does not limit its liability for death or personal injury caused by its negligence.

This licence agreement shall be governed by and interpreted and construed in accordance with English law.